Democracy in Danger

Democracy in Danger

How Hackers and Activists Exposed Fatal Flaws in the Election System

Jake Braun

ROWMAN & LITTLEFIELD
Lanham • Boulder • New York • London

Published by Rowman & Littlefield
An imprint of The Rowman & Littlefield Publishing Group, Inc.
4501 Forbes Boulevard, Suite 200, Lanham, Maryland 20706
www.rowman.com

6 Tinworth Street, London SE11 5AL, United Kingdom

Distributed by NATIONAL BOOK NETWORK

British Library Cataloguing in Publication Information Available

Library of Congress Cataloging-in-Publication Data

978-1-5381-2662-2 (cloth)
978-1-5381-2663-9 (electronic)

∞ ™ The paper used in this publication meets the minimum requirements of American National Standard for Information Sciences—Permanence of Paper for Printed Library Materials, ANSI/NISO Z39.48-1992.

Printed in the United States

For Jena, Alex, Cordi, Athena, and Hercules

Contents

Acknowledgments

I would like to thank my DEF CON Voting Village co-organizers Harri Hursti and Matt Blaze and DEF CON founder Jeff Moss for harnessing the collective power of the hacker community to improve election security. I would also like to acknowledge the tireless work on the frontlines of election security of dedicated election administrators like secretary Alex Padilla, Noah Praetz, Barb Byrum, Neal Kelley, John Odum, and others like them.

Thanks to Kendra Albert and the Cyberlaw Clinic at Harvard Law School for their sound advice to the Voting Village.

Thanks to Chris Burnham, Jane Holl Lute, and Doug Lute for their support and guidance. Thanks also to Jon Carson, Colin Bishopp, Laurie Moskowitz, Chris Button, Sherri Ramsay, John Rendon, and Phil Stupak.

Thank you to Dean Katherine Baicker and the University of Chicago.

I'm grateful to Diane Stockwell, my book agent, and Jon Sisk, my editor at Rowman & Littlefield Publishing Group, for bringing this project together.

Many thanks to Mary Hanley for her perseverance in research, fact-checking, and editing through many drafts.

Finally, I am most grateful to my wife Jena, our children Cordelia and Alex, and our dog Athena—thank you for putting up with this whole process.

Introduction

The Central Intelligence Agency (CIA), Federal Bureau of Investigation (FBI), and National Security Agency (NSA) have jointly stated with "high confidence" that the Russian government influenced the 2016 presidential election, with a dual aim of damaging Hillary Clinton's presidential campaign and undermining the U.S. democratic process. This book analyzes the historical context in which the attacks happened and their aftermath from a geopolitical, technical, and political perspective as it was lived by me and several hackers, political operatives, and national security experts whom I have lived and worked with for much of the last two decades.

Russia's interference in the 2016 U.S. presidential election provided an urgent and long-overdue wake-up call that U.S. election infrastructure is vulnerable to manipulation by foreign actors. As we approach the 2020 presidential race, election integrity has never been at greater risk; therefore, understanding how we got here and how we can solve the problem has never been more important.

This threat is new simply because there wasn't much election infrastructure to attack in previous generations. Only since the 2000 election debacle in Florida has the United States increasingly mechanized and networked its election administration operations. So while Russia has a long history of trying to affect U.S. elections, it would have been impossible for, say, the Soviet Union to have executed a similar attack during the Cold War since our elections were previously conducted almost entirely on paper, counted by hand, and results reported via telephone from one human to another.[1] And our election infrastructure is maintained with miniscule security budgets and operated by officials who have no background in technology, cybersecurity, or geopolitics. So without the help of the U.S. national security establishment, they are wholly incapable of protecting themselves from a cyberattack by a nation-state with the determination, sophistication, and resources of Russia. To be clear, hacking our election infrastructure and the mechanism that gives the citizen agency over her government is what this book is about. Not the fake news and social media morass that has been the topic of several other very well-researched books.

Moscow has clear policy objectives in hacking U.S. elections to keep the United States off-balance and focused on issues other than containing Russia. With the U.S. government mired in controversy concerning Russian attacks on the election infrastructure and looking inward, the Kremlin is free to shore up its sphere of influence in such places as Ukraine, Georgia, and Crimea. And Russia has a particular interest in promoting nationalism and splintering the European Union (EU) and North Atlantic Treaty Organization (NATO) alliances so it can engage the West on a bilateral nation-state basis where its bargaining power is stronger than negotiating with the EU or NATO as a bloc. In terms of his domestic politics, Vladimir Putin has successfully positioned himself as the only autocrat in the world who was bold enough to give U.S. democracy a black eye and get away with it. Moreover, since the 2016 election, we have learned that Russian election interference is no longer comprised of mere propaganda dissemination like so-called "fake news," but rather has graduated to the use of sophisticated cyberattacks aimed at the technological infrastructure used to administer elections.

Unfortunately, despite the clear policy objectives of the Kremlin and overwhelming evidence that election databases, networks, were breached, policymakers in the United States were paralyzed by a disconnect between an intelligence community who knew *something* had happened in the election and local election officials who have no experience with cybersecurity, saying they had "no proof that the election was tampered with,"[2] despite the fact that these election officials have nothing remotely approaching the technological capacity to know what hackers might have done to their systems. Suspicious of claims from local election officials that their systems were "air gapped" and "unhackable,"[3] white hat hackers at DEF CON, the world's largest hacker conference, felt compelled to test these claims and subsequently discovered gaping vulnerabilities in U.S. election equipment, networks, and practices.

These gaping vulnerabilities should have been no surprise to anyone overseeing voter protection on the Hillary Clinton campaign or the Democratic National Committee (DNC). The DNC, home to much of the institutional knowledge about the rickety infrastructure that holds our election administration equipment together, is particularly at fault since that is the institution, in the Democratic Party at least, that understands how nontechnical, low budget, and incapable the election officials would be in defending against a cyberthreat from a nation-state. Further complicating matters, members of the national security establishment have essentially no familiarity with voting infrastructure. This ignorance should have compelled them to ask tough questions when some election officials said they didn't need or want help defending our democracy from existential threats. The national security establishment should have exercised its mandate to protect our country from foreign attacks by insisting recalcitrant state and local governments take the free help they were offering so the American public could *at least* have confidence our nation's best cyberdefenders were protecting the sanctity of its votes. Unfortunately, none of these things happened. In the haze of a disaster, it is understandable that people didn't make the best possible choice at every turn; but what is less excusable is the continued denial of cybersecurity threats to our elections and refusal of federal dollars and cyber expertise in exchange for implementing common sense security practices by election officials. The national security establishment's continued refusal to fully exercise its mandate, to defend our nation from existential threats to the United States including a robust cyberdeterrence policy from the White House is also inexcusable. Fool me once, shame on you. Fool me twice, shame on . . .

For any Republican reading this who thinks, "Oh here we go again with another whining liberal blaming the GOP for disenfranchising voters," that is absolutely not the point of this book. In fact, time and again as I was working on voter protection issues for Democratic candidates, I was mortified to realize that more than 90 percent of the election officials standing in the way of Democrats being able to vote in a timely manner were Democratic election officials, not Republicans. Time and again, Democratic election officials would say things like, "I don't understand what your problem is, there have always been long lines here [in Philadelphia], what gives you the right to criticize us?" Or my favorite: "Voting is every American's duty, they should be happy to wait in line for hours to do it" (Dearborn, Michigan).

The reason I have experienced these issues is because, for better or for worse, I have worked on political campaigns, voter protection, and cybersecurity for more than a decade. I started my career as a campaign operative in Democratic politics in Chicago. I was loosely affiliated with Mayor Richard M. Daley's political organization and got to see up close the final hurrah of one of the last real urban political machines in the United States. I initially started as a journalist and then press secretary for several Daley-backed local candidates. My last press secretary job was for the first openly gay alderman in Chicago, Tom Tunney.

While I liked press and had a great time with Tom and other candidates I helped, I *loved* field organizing, largely because field is a massive logistics operation informed by evermore sophisticated algorithms. In a city as diverse as Chicago, where race politics is still a cornerstone of how campaigns are waged, I became captivated by conversations between the "bosses," in the true old-time sense of the word "boss," about how they would win a particular district. If the opposition was Polish and there were more Latinos in the district, then run an allied Latino candidate. What fascinated me even more was that these ruthless calculations of candidates' political fortunes were countable. This was in the early 2000s, before sophisticated data analytics was part of political campaigns. But with a city council district of 50,000 voters and only about 10,000 likely to turn out, we could easily figure out which voters were likely to vote in the election and then, between geography and first and last name searches, count up the Asian or white women in the district, or the Latino men or African Americans in the district (African American calculations are usually more based on census tract information). With some simple math and a little bit of voodoo, which I was told came from the gray beard bosses being able to just look at a ward map and "know" how many people would vote from each ethnic group, one could plan out who needed to get campaign mail, knocks on the door, phone calls, and so on.

Even more exhilarating, was witnessing this play out on Election Day. Let's say you look at voter history and assess that we need 1,500 more African Americans to turn out from previous elections and vote for our African American candidate against a white yuppie opposition candidate to win the election. So you decide to target 3,000 extra African American voters who didn't vote in the last election, but are likely to vote if we remind them repeatedly to do so. We decide that sending them five pieces of mail, three knocks on their door, and two phone calls would get half of those 3,000 voters to come out. Then, on Election Day, you find out that 1,532 more African Americans came out, and we won the election.

My mind was officially blown. What was also puzzling to me was that Chicago elections were in the dead of winter—usually February or March, when Chicago is at its coldest. When I naively inquired about this odd timing, I was told that years ago, the "real" machine bosses decided they could count on their patronage workers to get their friends and family out to vote come hell or high water because their city patronage jobs literally depended on it. This was the first time it dawned on me that the logistics of how or even when an election is administered affects the outcome.

After several years working in local Chicago politics, I tried to move into national political circles by working on John Kerry's presidential campaign in Iowa. Iowa campaigns forgo Chicago's Machiavellian urban machine race politics. However, they are actually more scientific and technology savvy than almost any urban area in the country. Frankly, if one wants to know about the most cutting-edge tech and data analytics going on in a campaign at any given time, don't waste your time in Silicon Valley, go to Iowa. In Iowa there isn't a "political machine," and no one is purposely trying to make sure people can't vote; but even in Iowa, how an election is administered is crucially important. Multiple candidates like John Kerry and Barack Obama won the Democratic nomination in part because Iowa's presidential nomination process is administered as a caucus. Not a typical primary election. Fortunately, for me, the election administration process had worked in my favor up until this point—meaning election administration procedures were contributing to my wins. That didn't last though. In Michigan in the general election of 2004, I saw the dystopian hellscape that is poorly administered elections. This version of elections only works in favor of those who don't want people to vote.

As I will go into later, elections in Detroit and many other places were managed terribly in 2004. It was particularly bad in places where minorities and students lived. Lines at the polling place stretched for hours and blocks, keeping thousands of students and minorities from voting. In Ohio, mismanagement led to so many people voting provisionally and such long lines it may have cost Kerry the election. Several political operatives in 2004 vowed that we would not lose the next election because of how it was administered. We took nothing for granted in terms of election administration in 2008. A team led by me and the voter protection department dug into every aspect of how elections are administered in this country—everything from how long it takes to vote on each type of machine in the country, to which queuing process creates the worst bottleneck, to how long it takes to enter each new voter registration into the voter registration database, to, literally, the number of parking spaces available at peak voting times in a polling place. We knew the ins and outs of election administration logistics so well that we even built an algorithm tool that, with the appropriate inputs, accurately predicted wait times at a given polling place. Nothing like it had ever been created in the election industry, and it is now housed on a public nonpartisan platform at the Massachusetts Institute of Technology (MIT)[4] and used by election officials throughout the nation.

But, after 2008, I stopped paying attention to voting administration issues because I was asked to serve as deputy director of the Presidential Transition Team for the National Security Agencies Review for the Obama–Biden transition. In that capacity, I oversaw logistics for the review of the national security agencies—the Department of Defense, the State Department, the Department of Homeland Security (DHS), the FBI, the CIA, and so on. It was in this role I learned, firsthand, the immense challenge cyberattacks posed to our national security. Once Obama took office, I was asked to serve as White House liaison to DHS.

At the beginning of our tenure at DHS, the department was required by Congress to complete the Quadrennial Homeland Security Review (QHSR), which was a review of the agency to be delivered for congressional oversight.

Between the review of DHS for the Presidential Transition Team and now an in-depth review of DHS for Congress, I felt like I knew the agency inside and out. But throughout these reviews, the topic of cybersecurity consistently seemed out of place. It was clearly an important issue; however, as far as the civilian agencies were concerned, it was largely relegated to the IT department. In fact, until about two weeks before DHS was supposed to deliver its QHSR report to Congress, cyber wasn't even listed as one of its top five policy priorities. After a little drama between the cyberdivision, the Office of Policy, and the front office, it was inserted as one of the top five priorities. Prior to that time, cybersecurity was not listed in any official DHS documents as a top policy priority.

This opened up an opportunity. Now that cyber was a top policy priority, people in the front office, like myself, could request briefings on the agency's cybersecurity strategy and programs. After about a year, the deputy cabinet secretary, Jane Holl Lute, was asked to delve into "cyber policy," as it was now called. For the first time, she dedicated a small section of the Office of Policy to cyber policy analysis.[5] Fortunately, I had been tasked with working for Deputy Secretary Jane Lute to pass a big data-sharing agreement between the EU and the United States. We were sitting in a hotel lounge in Brussels having dinner and a drink after a long day of talks with EU officials. Deputy Secretary Lute had just gotten word that one of the final stages of the agreement had passed through the EU. So we were all breathing a sigh of relief. She looked at me and said, "Well now that this is almost over, you want to help me figure out what we are going to do about cyber?" I almost jumped out of my seat and hugged her, as I had been wanting to work directly on cybersecurity for some time. Instead I tried to hold it together and said something along the lines of, "Yes, of course, ma'am. I am more than happy to help."

From that moment onward, 90 percent of anything I did for DHS or really in the national security space in general was in some way tied to cyber. I also got a front row seat to see some of the most dramatic cyber policy fights in the federal government play out. Lute was at the center of a fight between DHS, NSA, and the White House about who would oversee domestic cybersecurity. This was a fight that, thank goodness, she won. It may seem anathema to us now, but before Edward Snowden there were serious debates among incredibly intelligent and reasonable people as to whether NSA should quarterback our civilian cybersecurity. From a purely technical proficiency standpoint, NSA had a point, given that they have better tools than anyone else. From a constitutional perspective, it was lunacy. Thankfully Jane won the day. Had she not, the Snowden revelations regarding data collection on U.S. citizens would have been far worse.

I also got a front row seat for the struggle to define and implement basic cybersecurity on federal government networks. Until 2012 or so, there was little to no consensus on what basic cyber "hygiene" entailed—the equivalent of a network brushing its teeth, flossing, and washing its hands regularly. This time NSA was quite helpful, as they had defined the most basic "cyber hygiene" practices an entity must implement to protect itself from nefarious actors.[6] As one can imagine, no agency wanted to be directed by DHS regarding its own cybersecurity, and the White House was cautious at best at giving DHS control over implementing such cyber hygiene. Yet, Jane won again, and now DHS not only oversees implementation of basic cyber hygiene for the civilian agencies, but also has "hunt" teams to go inside agency networks to root out bad guys.

While it may be the most mundane aspect of what I did regarding cybersecurity at DHS, the project that gave me the most useful perspective on cybersecurity at DHS was helping run a task force on cyber workforce development. In short, we were trying to figure out how prepared our cyber workforce was to complete their mission. Also, we assessed how to replace workers who left or train existing staff on the constantly evolving skills required to perform the mission. We knew DHS couldn't pay what the New York financial industry or Silicon Valley tech industry pays for cybersecurity talent; but we had several things going in our favor. We had meaningful work to do protecting the country from cyberattack that would give anyone a reason to get out of bed in the morning. The work was also incredibly cool and sexy. Much of it you would get arrested for doing anywhere else. Also, while the pay wasn't the best, it was stable, with good benefits and guaranteed vacation.

What we found was jaw-dropping. DHS didn't even know which of its positions should be categorized as cybersecurity positions. So, it had no idea if the people in those positions were proficient at cybersecurity. There was *one* federal employee assigned to recruit top cyber talent for the entire agency. There was no program in assessing cyber workforces' technical capability. DHS didn't even have a list of the basic cyber skills a worker needed to achieve her mission. Training for new skills was ad hoc and usually up to the employee. If she was lucky, tuition assistance was available. At the outset of the cyber task force, we thought our findings would send us to the marquee cyber competitions and hacker conferences, job offers in hand, to recruit the best and brightest to work at DHS. Instead, the task force illuminated that we first needed to *count* which of the 250,000 jobs at DHS were cyber jobs and which of those were unfilled. Once that was done, we realized our human resources department couldn't even generate an accurate job posting to solicit applicants with skills germane to the DHS mission. This problem arose because no one in human resources knew anything about cybersecurity. How could we expect HR to write a job posting on something they knew nothing about? So that was our next mundane task: build a tool to automate job postings for specific cyber skills. The cyber task force made implementing the sexy world of cybersecurity as unsexy as one can imagine.

For me though, the crucial knowledge acquired from this project was less about HR practices than realizing DHS could not hire the talent it needed to fully execute its mission because there simply weren't enough people with the requisite talent available. It dawned on me that if DHS was in this much of a mess, everyone else besides the military and most tech savvy corporations must have essentially no cybersecurity at all.

It turns out the workforce shortage is not unique to DHS. Depending on whose estimates you believe, there is anywhere from a 200,000-person to 800,000-person shortage in cyber workers nationwide.[7] A shortage in teachers, for example, means larger class sizes; but fewer cyber professionals means many, if not most, organizations can't hire anyone at all. If they can't pay top dollar, like the banks, or offer incredibly cool work, like NSA, they are largely out of luck. Even if an organization does get lucky and find a cyber expert to work for them, the person is usually poached within a year or so by one of the more appealing organizations.

I saw this firsthand after I was appointed to the Illinois governor's cyber task force. It was really called the Internet Privacy Task Force,[8] but I convinced the staff on Governor Pat Quinn's team to let the task force focus on cybersecurity more broadly rather than just privacy. In that capacity, I worked with multiple state officials, the likes of state Chief Information Security Officer Rafael Diaz. Rafael and his team are incredibly smart and talented but hamstrung by lack of qualified cyber personnel. In fact, no recommendation made by the task force was taken seriously if it required cyber professionals to implement because there was none available. As I spoke to more and more state and local cyber professionals, I began to realize this was a problem at near-epidemic levels. They desperately needed to find competent people.

It was clear that only a handful of people are minding the store in state and local government or private-sector firms. And of those who are minding the store, many of them are not cyber experts at all. Again, this isn't their fault, there is no one to hire.

It will become clear later in the book that this human capital issue has an outsized impact on election security. If the chief information security officer for the state, generally the top cyber position, struggles to hire cyber experts, imagine how hard it is for the secretary of state or county clerk who oversees elections to hire anyone competent enough to implement good cyber hygiene. Unfortunately, we found that election officials turn to those they think will have answers: the vendors and election industry lobbyists in D.C. Neither of these groups know much about cyber either. And they are conflicted at best in providing correct answers to these technical questions.

Shortly after serving on the Illinois governor's task force, I was asked to teach a class on cyber policy and later election security at the University of Chicago's Harris School of Public Policy. The school later asked me to create its first ever cyber policy initiative, focusing on education, research, and cyber public policy development.

Where these three areas converge—politics, election administration, and cyber policy—is where this book begins and ends. While I neither created nor planned to be in the scrum of all this, I certainly find it fascinating. The newness of it all makes it even more interesting to me. The term "cybersecurity" wasn't a thing fifty years ago. The term "cyber policy" wasn't a thing even 10 years ago. Now "cyber politics" isn't even really a thing yet; however, I contend that cyber politics will increasingly become central to our lives, whether we realize it or not. That is, the politicization of cybersecurity and the injection of cyber into our political discourse is here to stay and will only grow in coming years.

During my time at DHS, Deputy Secretary Lute repeated two lines that stuck with me. The words lodged themselves in my frontal cortex to be deconstructed and reassembled over and over again. The first line was, "It took 20 years for the first 4 billion people to get connected to the internet. It won't take another 20 years for the last 2 billion to be connected." The other line was, "Once it [cyber] is in the paint, we know things have radically changed."

While she always said the two lines separately, they are really part of the same macro trend. The Fourth Industrial Revolution involves connecting the last 3 billion people and everything humans make (including paint) to the internet. Connecting everyone on the planet and everything people make to the internet will happen in our lifetimes. Once it happens it will likely never be undone. New people will be born and new things made, but in the future, the default will be to immediately connect them to the internet. A thousand years from now this ubiquitous connectivity will be a moment historians rank alongside the creation of agriculture, the printing press, and electricity as a major inflection point in human history, after which nothing is wholly separate from it and everything is in some way referential to it.

This march toward universal connectivity leads us to where this book ends. That is the uncomfortable and undefined world of cyber politics. Once we have achieved ubiquitous connectivity and the internet is inextricably linked to our society, it will inherently become political. As Aristotle and most of the other classical Greek thinkers, as well as Jefferson and the leaders of the Enlightenment, stated, "Humans are political animals."[9] If Aristotle and Jefferson are right, that what is human is political, then it must follow that as the world of cyber becomes more directly tied to our humanity, it must also become inherently political.

The politicization of cyber is gut-wrenching to those who understand cybersecurity. It is discombobulating to political operatives. Everyone else just feels like they are living in a Technicolor version of *Leave It to Beaver.* Things kind of look like they used to, except with new lenses that enhance your perspective, but it still seems off. Putin understands this. Trump likely does, too. I am just not sure anyone else does.

• 1 •

Table Setting

The Putin Problem

Obama was wrong. I remember laughing watching the presidential debate in 2012, between President Obama and Governor Mitt Romney. Romney had said in a previous interview that Vladimir Putin's Russia was the United States' "number one geopolitical foe."[1] Obama questioned him on that statement in the debate, quipping, "The 1980s are now calling to ask for their foreign policy back, because the Cold War's been over for 20 years."[2] In fairness, I completely agree with the rest of Obama's foreign policy, including the so-called "Pivot to Asia," getting out of Iraq, reducing our military commitments in the Middle East and Central Asia, the Iran nuclear deal, and the Paris Agreement on climate. But, on this one, Romney was right and Obama was wrong. Putin is a grave threat to the United States. The main reason I didn't think I should question Obama was that I am no Russia expert. And even the hawkish George W. Bush had famously said, "I looked the man [Putin] in the eye; I found him to be very straightforward and trustworthy. We had a very good dialogue. I was able to get a sense of his soul, a man deeply committed to his country and the best interests of his country."[3] Needless to say, Bush and Obama differed greatly on foreign policy. So, if they both agreed on Putin and Russia, who was I to disagree?

But after speaking with dozens of national security leaders and reading thousands of pages on the topic since the 2016 Russian attacks on our democracy, it is clear Russian experts are in agreement that Putin's Russia is a grave threat to the United States, specifically our democracy.[4] The best example of their consensus may be the unclassified version of the National Security Agency (NSA), Central Intelligence Agency (CIA), and Federal Bureau of Investigation (FBI) joint report on the 2016 election attacks.[5] In fact, some national security experts argue the 2016 attacks could eventually be *more* damaging than 9/11 or Pearl Harbor.[6] That seemingly hyperbolic claim is made by these traditionally measured national security leaders because while the loss of life and property was tragic and devastating in 1941 and 2001, we are a resilient nation and were able to bounce back. More than just bounce back, we came out of those tragedies swinging in ways that annihilated our enemies and even reshaped geopolitics as we know it.

Conversely, the 2016 attack killed no one and destroyed no physical property that we know of. Rather, it struck at the core of every American's relationship to our government: the integrity of the ballot. It placed at risk our trust in the outcome of the election, and more than just one election, it began to undermine public trust in our democracy. Beyond Donald Trump and, to some degree, Hillary Clinton claiming the election in 2016 was "rigged," we have already seen this trust continue to erode in the subsequent 2018 midterms as candidates from both parties in Georgia and Florida made claims that the democratic process was "rigged" or "hacked."[7]

Attacking this ethereal belief that our democracy is both free from corruption and fair for both voters and candidates is a larger danger to our nation. As the most recent former U.S. ambassador to the North Atlantic Treaty Organization (NATO), General Douglas Lute told me, "Putin has decided he doesn't have to beat us with conventional means, tanks and guns. So, he is going to try to erode our democracy from within." He is attacking our civil institutions, first among them being the democratic election infrastructure. He is further exploiting existing fissures within our society, like those between the left and the right in the country.

In short, Russia's cyberattacks on our voting infrastructure are a national security threat to the United States and its allies. It is not simply conjecture or rhetoric to say the 2016 attacks were a national security threat. The attacks themselves fit the definition of a true "threat" in national security parlance; Russia has both the capability and intent to attack.

In a conversation with Ambassador Lute, he explained the national security community perspective with five key criteria:[8]

> First, these election attacks are a national security issue because Putin has already executed a successful attack against our elections with cyberweapons. This isn't hypothetical. It's something he has already done. Second, election security is a national security issue because Russia is here to stay. Putin will be in power until 2024 and perhaps longer. This is not a one-shot deal where Putin decided he didn't like Hillary Clinton and decided to hack our elections. Third, this is a national security issue because other potential adversaries are watching and devising ways to further undermine our democracy. The 2016 attack could inspire copycats, for example perhaps China, Iran and North Korea. Fourth,

> this is a national security issue because the opportunity for Russia to attack again is right around the corner in 2020. We know they probed the election systems in about one-half of our states in 2016 and did modest attacks in 2018.[9] They undoubtedly learned from these attacks and now understand our systems better, meaning that we should expect in 2020 more focused, sophisticated attacks that are harder to attribute to Russia. The final reason this is a national security issue is that our allies are vulnerable as well. The Russians have been engaged in election interference, even among some of our closest allies, for years before 2016.[10] And after Russian hackers penetrated the U.S. election infrastructure in 2016, we saw subsequent attacks in France, Germany, and it [was] reported in the United States again in 2018,[11] this time in Knox County, Tennessee.[12] Brexit, for example, has thrown the United Kingdom into nearly three years of gridlock. Russia's attacks on the Brexit referendum are widely believed to have had some impact on the outcome.[13] It only serves to weaken the United States hand abroad when our best allies are made weak as well.

A key question one must ask here is, "Why?" Why would Putin take such risks to weaken the United States and its allies? What happened to the late 1990s Russia that was transitioning toward democracy and a market-based economy, as favored by most of the developed world?

Again, I turned to former U.S. ambassador to NATO Douglas Lute. His answer was as striking as it was concise: "Very simply, Putin has one overriding objective: to stay in power. But he knows he has a weak hand in the long run. Russia is a state in steady decline: Economics, health, demographics, and political-social indicators are all in decline." To overcome these challenges, Putin must keep potential rivals close like the military, intel services, and oligarchs. So he points to external enemies, like the West and especially the United States, to align with these power brokers. Furthermore, he suppresses domestic opposition and stokes nationalism by emphasizing the threat of external enemies that justify his authoritarian grip on power.[14]

To counteract this perceived external threat, he is doing what every Russian czar since the Mongol invasion of the thirteenth century has attempted to do, which is to fortify Russia's position with a buffer zone of friendly or at least weak and compliant states along his borders.[15] So what are the threats to this "buffer zone"? Encroachment of the European Union (EU) and NATO. The EU and NATO represent the two most powerful multilateral institutions promulgated by democracies of the West. Indeed, they are the most powerful multilateral institutions both economically and militarily on the planet.

Here the West deserves some criticism for not seeing this coming. If Russia's economy were booming and a leader committed to democratic institutions, like Boris Yeltsin, were in power, we could have likely expanded the EU and NATO right to Russia's doorstep with less concern for the consequences.

Yeltsin's Russia may have even seen the EU and NATO expansion as less of a threat because the EU would have made trade easier with the myriad states in the region, and NATO could be sharing the burden of fighting the various terrorist organizations operating in Eastern Europe and Central Asia. But with a weak domestic economy and such popular movements throughout as the color revolutions in the former Soviet space and the Arab Spring challenging dictators, Putin was bound to respond brutally to EU and NATO advances toward Russia's borders. This is why he reacted violently to Georgia's flirtation with the EU, invaded Crimea, and continues to destabilize eastern Ukraine.[16]

As Putin needs to create an enemy Russians can rally against, he is compelled to oppose EU and NATO expansion. Consequently, he annexes Crimea, a long-standing Russian prize, and is hit with sanctions from the West in return. The sanctions further weaken his already weak economy.[17] This starts a vicious cycle where Putin will exploit any fissure in the West to divide us against ourselves and further enhance his buffer zone. Nonetheless, Putin's challenge is that he doesn't want to spark a shooting war or trigger further sanctions. So, he must use weapons that are hard to attribute. And these weapons need to inflict the maximum damage while operating just below attacks that would trigger a war.

Enter the "perfect weapon," as David Sanger, chief national security correspondent for the *New York Times*, deems cyberweapons in his new book with the same title. Cyberattacks are hard to attribute to a particular attacker, and as long as it doesn't result in loss of life, they live in a gray area between espionage, military, and propaganda attacks, unlikely to invoke a military response. Moreover, these cyberweapons are useful because of the target. As stated earlier, Putin is not capable of destroying our military installations and making us succumb to his will without a massive counterattack from the West. But cyberattacks allow him to exploit fissures to divide us via both nationalist movements on the right and anticapitalist or race-based movements on the left, and at the same time undermine the democratic institutions that enable us to sort out these differences through free and fair elections. Cyberweapons further enable him to keep attacking us because even when he gets caught, there is no "smoking gun." So he can claim it wasn't Russia even though every rational person knows it was Russia, and Russia knows every rational person knows it was Russia, yet no smoking gun. No Pearl Harbor. No Twin Towers. This discombobulation confounds a reasoned and measured response from his adversaries even more.

This being said, none of these dynamics, in principle, are new. This is just the most recent example of an asymmetric weapon being used by a weaker state against a stronger state, or in this case a stronger state (the United States) and stronger multilateral institutions (the EU and NATO). In fact, whether cyberweapons are the perfect weapon or, more specifically, the perfect asymmetric weapon is yet to be determined.

But what does seem to be new and menacing is the extent to which cyberweapons can be used during peacetime to achieve political ends far beyond what peacetime measures were previously able to achieve between great powers. For example, during the so-called "Great Game" of seventeenth-century peacetime between the British Empire and the Russian Empire regarding Central Asia,[18] diplomatic and propaganda efforts often gave one or the other nation a perceived upper hand in the region; however, when either side was determined to exert control over geographic boundaries, it deployed its military in the form of hundreds or thousands of troops. In contrast, Putin has been able to attack the United States' democratic institutions and stoke the flames of division between the extremes of both parties without sending any troops to the country. He has also been able to pit the president against his own intelligence agencies and instigate the gravest internal challenge to a president of the United States in almost a half-century. Despite the report being completed, the FBI Special Counsel investigation led by Robert Mueller regarding whether the Trump campaign colluded with Russia to affect the 2016 election or subsequently attempted to obstruct justice is still gridlocking Washington.

Gone from the headlines are stories about Russia's newest incursions into Ukraine or Crimea, or support for Assad in Syria, much less his newest ventures in Moldova,[19] Montenegro,[20] and Azerbaijan.[21] The sanctions passed by the Senate for interfering with our elections have yet to be fully implemented.[22] Instead, headlines related to national security are dominated by Trump, Mueller, and Russia's involvement in the election. It may be the greatest smokescreen ever behind which Putin can meddle and rebuild his buffer zone.

While I contend that using cyberattacks to influence politics is relatively new, this is not the first time a state has used cyber to influence politics and bide time. In fact, the most well-known example of a political cyberattack originated in the United States, with Stuxnet, as the NSA Olympic Games program has come to be known, a joint attack by the United States and Israel on the Iranian nuclear program. The story of Stuxnet has been well-documented in the media, as well as in books by David Sanger and Kim Zetter,[23] and a film entitled *Zero Days*. So I won't go into it too much here. Essentially, the Bush administration was feeling pressure by the Israelis to bomb the Iranian nuclear program but decided, rightly so, that it could not start yet a third war against a Muslim country. The Bush team came up with a cyberattack that would degrade the nuclear program's capability while stopping short of what they believed would be considered by the Iranians as an act of war.

Obama kept this program going during his tenure and seems to have expanded it. The technology behind Stuxnet is fascinating; however, the shortened version is that this digital worm needed to get into the Natanz nuclear facility in Iran, a heavily guarded military installation in the middle of the desert, most of which was buried underground. Once inside, the worm would need to cross an "air gap," which means it would need to jump from one network to another without those networks being connected digitally. Once inside the network of the nuclear program, the worm either sped up or slowed down the machines that make the nuclear material, called centrifuges, causing them to malfunction and ultimately self-destruct. The fact that NSA and Israel's Mossad could pull this off is remarkable in and of itself. In fact, in the film *Zero Day*, NSA officials are quoted as boasting that they "always found a way to get across the air gap." The NSA team actually had to "jump" the air gap multiple times during the several-year tenure of Stuxnet. "Jumping the air gap" means the NSA hackers had to not only get the worm into the concrete vaults buried underground in the desert but once inside the worm would have to "jump" from the regular computer network in the plant on to a completely separate, unconnected network the nuclear centrifuges were on. Getting Stuxnet into the centrifuge system was just a first step. After that, updates needed to be sent to the weapon, new commands, and so forth, and each one of these modifications had to jump the air gap. They all did.

Even more interesting was that the NSA team never destroyed all the Iranian centrifuges at once. They only destroyed about 10 percent of them at any given time—far higher than the normal rate of wear and tear but low enough that one could never be sure that they were being attacked. So we were attacking their centrifuges for one or all of three reasons: (1) to get them to not trust their own technical capability to build a bomb, (2) to make them so paranoid that the United States or Israel was in their networks that they would never get the bomb, or (3) to simply buy time, to set the program back sufficiently enough to buy time to negotiate.[24]

The attack wasn't really aimed at the nuclear program, but rather the Iranian political elite. It was designed to put them in a state of decision-making paralysis. Are their scientists not good enough to build a bomb? Were they being sabotaged? If so, by whom? Would getting a bomb cost far more than they had initially projected? While they and their top engineers undoubtedly were sent into a state of total paranoia, the United States and its allies had years to get them back to the negotiating table. Frankly, this worked. Among other things, we likely would not have gotten a nuclear deal with Iran had Stuxnet not been deployed. Maybe getting caught was the goal all along.

What's more interesting is that Stuxnet was first discovered by the international cybersecurity community and then the Iranians a couple of years before the nuclear deal was signed. By that time, the Iranians had purged Stuxnet from their systems. It is believed an even more sophisticated cyberweapon than Stuxnet had gotten past the "air gap" yet again; however, the initial cyberweapon had been removed, and yet the Iranians were still negotiating. Why? Had they not beaten the vaunted NSA and Mossad? Not if the target was the Iranian political elite.

While several observers disagree with me on this point, I contend the final "payload" delivered in this political cyberattack was getting caught. It's one thing to slow the nuclear program and make the Iranian elite paranoid. It is quite another to achieve this and then have the Iranians find out that the NSA has, in fact, been operating in their "air-gapped" vaults buried in the desert for years without their knowledge. Obviously, this attack was done with several-year-old cyberweapons. So they would be constantly haunted trying to imagine what the NSA must be currently hacking them with. The attack doesn't achieve its full effect, then, until the target of the attack knows they were owned by the attacker, especially when the attacked was owned for a time without their knowledge. Evoking these feelings of helplessness is a political act in nature, especially as it was designed specifically to affect the Iranian political elite, not the general population. It's possible that getting caught was unintentional, but it was far more humiliating for the Iranian political elite than it was for the United States and Israel to have Stuxnet outed.

Here again is why cyberweapons prove to be the perfect asymmetric weapon for a weak state. The cost–benefit ratio is in the benefit of the attacker.[25] The United States and Israel undoubtedly spent billions to produce Stuxnet and its subsequent iterations. But a state could achieve similar ends with far less funding. For example, Russia inflicted a cyberattack on the Estonian government for likely a few million dollars.[26] The attack was also a political response to a perceived Estonian attempt to distance itself from Russia; however, it was essentially a DDoS (Distributed Denial of Service) attack on Estonian government websites.[27] These DDoS attacks made the sites inaccessible for Estonians wanting to access basic government services. In Estonia, this type of attack had an outsized effect because it is an e-government in which all government services are accessed online. So shutting down government websites brings the country to a halt. Aside from being inexpensive, the cyberweapons used in these attacks adhered to the traditional definition of an asymmetric weapon because they had a level of anonymity that enabled the attacker to deny involvement and, hence, attack another day.[28]

These asymmetric weapons have an outsized impact compared to their low cost. Like a roadside bomb that costs about $40 to make but can take out a $7.5 million Abrams tank,[29] the cyberweapons used to take out the entire Estonian government network and bring the country to a standstill by shutting down the Ukrainian electrical grid cost little to develop and dispatch relative to the damage they inflicted on their targets. Possibly the most insidious aspect of asymmetric weapons is that they are incredibly hard to defend against. Like a roadside bomb that can be disguised in myriad ways, cyberweapons are similarly hard to defend against simply because our technology is designed for speed and ease of use, not security. For example, 46 U.S. banks' websites were attacked by the Iranian Revolutionary Guards as a response to Stuxnet but were never able to fully defend themselves, despite spending billions on security.[30] The attacks only stopped when the Iranians decided to end them, after more than *two years* of successful attacks.[31] As Douglas Lute explained, "Asymmetric weapons have three characteristics: a cost–benefit ratio in favor of the attacker, very hard to defend against, and anonymity. Cyberattacks are the twenty-first-century version."[32]

So what we have is a shrewd despot, Putin, sitting atop a declining state, Russia, determined to hold on to power by using the United States and its allies in the West as enemies his citizens can rally against while rebuilding some form of the Soviet or czarist buffer states around his country. The despot has astutely reasoned that he cannot outmatch the United States et al. with conventional military forces. He has decided the most expedient course of action therefore is to make our societies crumble from within, through first exploiting existing fissures within the populace and then undermining our central tool to sort out those differences: free and fair democratic elections. His preferred tool in this ambitious endeavor is an asymmetric cyberweapon. Like any good asymmetric weapon, this one has a cost–benefit ratio in his favor and is so inexpensive he essentially has unlimited resources to develop and deploy the weapon. Once deployed, he can avert retaliation by denying deployment of the weapon because it is quasi-anonymous once delivered. Finally, he is assured some level of success with each attack because the cyberweapon is incredibly hard to defend against.

We know that this is not academic pondering but a clear national security threat because he has deployed the threat and plans to deploy it again imminently against the United States and our allies. He has already successfully deployed this playbook against Estonia, the Ukraine, and the United States, among others, causing the desired degree of domestic political chaos and foreign policy gridlock. And we know this playbook works because we have deployed it against Iran, to great success as well. Cyber is proving to be an effective asymmetric geopolitical weapon. What seems to be new is that it is seeping into partisan electoral politics as well.

It seems Putin is holding several cards. But the United States had been safeguarding elections domestically for more than 200 years. Surely he will be surprised by key defenses we have in store for him that he didn't anticipate. Or not.

• 2 •

The Line to Nowhere

Fourteen years earlier . . .

"What the fuck do you mean there are going to be long lines?"

It was afternoon in late October 2004, just a few days before the general presidential election between George W. Bush and John Kerry, and I was on the phone with Laurie Moskowitz, who was overseeing Get Out the Vote (GOTV) for the Democratic National Committee (DNC). As we were talking, she had nonchalantly asked, "Oh by the way, what's your long line program?"

At the time, I was serving as the statewide field director for John Kerry's Michigan battleground state operation. For the first four months on the ground, it was my job to manage more than 100 field organizers who were recruiting and training volunteers to talk to undecided voters and persuade them to vote for Senator Kerry on Election Day; but in the last six weeks, my job had increasingly been to turn that army of volunteers and staff into an organization that could best Karl Rove's vaunted "72 Hour Program," a method, perfected in 2000, that was meant to overcome the disadvantage GOP candidates believed they faced from organized labor and the remnants of urban machine operations in turning out voters on Election Day. Democrats, by contrast, were evolving their GOTV operations to compensate for the declining power of organized labor and the disappearance of urban machine capability to turn out voters who were unlikely to vote. Unlike the old machine days of such powerful urban mayors as Daley, Pendergast, and Tammany Hall, modern day big-city mayors and political bosses were largely incapable of running an effective GOTV program. The Democratic Party's inability to turn out voters on Election Day was causing political observers to refer to Democrats after the 2000 election with terms like "permanent minority."

It was specifically this GOTV production gap between Democrats and the GOP that I, along with the other fifteen or so battleground state field directors, was hired to overcome. Our primary task was to reverse the historic downward trend of the Democrats' declining ability to turn out voters, coupled with recent GOP proficiency in doing so. Thus, anything that might hinder my team's GOTV efforts, for instance, long lines to vote on Election Day, was particularly disconcerting to me. Under the direction of longtime Democratic field guru Michael Whouley, I had developed a nuanced understanding of these voters who were unlikely to vote on Election Day. We spent an enormous amount of time reading polls, trying to figure out why they didn't vote, calling them on the phone and asking them what issues would motivate them to vote, and knocking on their doors and asking them what might stop them from voting on Election Day. In fact, my back-of-the-envelope math shows that I, personally, likely spoke to 27,000 voters on the topic of voting on the phone, on their doorsteps, or at rallies in the eighteen months leading up to the 2004 general election.

The citizens who don't often vote tend to be young people who don't yet fully grasp the importance of voting, single mothers who are busy juggling kids and multiple jobs, or disaffected low-income people who feel government isn't working for them or society is leaving them behind and therefore don't see the point in voting. For Republicans, these voters were in low-income housing in the exurbs. For Democrats, these people were either in college towns or low-income housing in cities.

Since I had gotten to know so much about these low-propensity voters in the last few months, I was confident they shared one trait: They are not the type of people who are going to wait in line very long to vote, especially if they had to wait outside in freezing November temperatures. In many cases, these people had never voted before and had already heard it was a hassle. I thought the young people would lose interest and leave to find something more entertaining to do. The single moms often have kids with them who are not going to wait patiently and invariably need bathrooms and food. The disaffected and poor voters usually have service industry jobs where arriving more than 15 minutes late leads to serious consequences. These concerns led to my second question.

"How long are the lines?" I asked Laurie.

"Depends." she said. "Sometimes they can get up to five to six hours long."

"Whaaaaaaat!?!?!? Are you fucking kidding me?!?!!? I thought you were going to say 30 to 45 minutes. These voters aren't going to wait around for six fucking hours to vote! I fucking wouldn't wait for six hours to vote, and this is my fucking job! We are going to lose thousands, maybe tens of thousands of votes!"

My team in Michigan had done an exemplary job so far in its GOTV efforts, but I was still nervous. The polls were tightening significantly. After Bush's dismal first debate performance, he rebounded with a strong showing in the last debate. Even more worrying was a new video from Osama bin Laden that was shown over and over by the media. We knew that anytime bin Laden's face was on TV, we were losing votes. Kerry had been up as much as 5 percent over Bush in Michigan, but polls in the closing days showed us tied or only 1 percent ahead.[1] Things were definitely going in the wrong direction. We could not afford to lose thousands of votes due to the fact that our voters wouldn't or couldn't wait in long lines. This obviously led to my next question:

"Well what the fuck are we supposed to do? How do you deal with this? Why the fuck didn't anybody tell me there were going to be long lines on Election Day?!!? We should have been figuring out how to address this for the last four months, not the last four days!"

Laurie was a little taken aback. "Why didn't you know there were going to be long lines?" she asked. "There are always long lines on Election Day."

I had no idea that long lines to vote are a reality for many American voters. I am from Nebraska, in a neighborhood in Omaha called Prairie Lane, which is much like most American suburbs. I don't remember going with my parents to vote when I was a kid. My parents, faithful Republicans, take their civic duty seriously and always vote. The most likely reason I don't remember going with them to vote is that it was totally uneventful. Omaha may not be the most exciting place on earth, but it runs efficiently. I had started working in politics in Chicago, but I didn't start working as a paid staffer on campaigns in the city until 2001, after the 2000 election. Elections in Chicago, where I began my career as a campaign staffer in 2001, are similarly well run, largely thanks to the efforts of Chicago's then-city election commissioner, Lance Gough and in neighboring Cook County by David Orr and his senior advisor, Noah Praetz. Furthermore, turnout is generally lower in nonbattleground versus battleground states in presidential years because the two political parties are not spending millions of dollars to turn out voters on Election Day.

The first presidential election in which I was eligible to vote was the 1996 presidential election. At the time, I was in a study abroad program in Rome, and my absentee ballot arrived after Election Day. That should have been my first clue that elections were poorly administered in the United States, but I chalked it up to bad luck and forgot about it. In the 2000 election, I voted early in Chicago, days before Election Day, and didn't notice any long lines.

It's unclear how long this issue of several hour-long lines has been a problem in major urban areas. As I will explain later, the number-one cause of long lines in recent years has been touchscreen voting machines; however, touchscreen voting machines weren't introduced on a large scale in the United States until after the Help America Vote Act (HAVA) was passed in 2002.[2] Until then, many jurisdictions used punch cards to vote,[3] a process made famous by the Florida vote recount debacle in 2000. There were places like Wisconsin that always had long lines in their urban areas because of a particularly inefficient yet statutorily mandated process for checking people in at the polling place; but by and large, the voting process was pretty straightforward, and while people often had to wait to vote, long lines weren't an unsolvable problem on Election Day the way it is unsolvable with touch screen machines, which I will explain later. But after the $3.2 billion HAVA bill was passed, local jurisdictions went on a buying spree to get equipment that would ensure the term "dangling chad" became a thing of the past. This new equipment would have unintended ripple effects for countless voters.

Fast-forward to the first election, where touchscreen voting machines were widespread and it soon became clear that many election administration officials had not considered what is, in essence, a simple math problem. It should take all of two minutes to calculate the results if you have five touchscreen voting machines in a precinct and it takes six minutes for the average person to cast his or her vote, multiplied by thirteen hours (the polls in most states are open for twelve or thirteen hours, most commonly from 7:00 a.m. until 7:00 or 8:00 p.m.),[4] you can expect exactly 650 people to cast their votes in that precinct on Election Day. Period. So if you expect 1,000 people to vote in that precinct, you have a big problem: either you have to keep the polls open for another three and a half hours or you have to disenfranchise the 350 people who you will turn away from the polling place when the polls close at 7:00 p.m.

This arithmetic would be unacceptable even if there was a steady stream of voters throughout the day; but more than a third of voters show up to vote after work, between the hours of 5:00 to 7:00 p.m. Moving 333 people through the voting process at six minutes a person on five machines takes almost seven hours. So, voilà, this is very simply how you get a six-hour line.

Of course, this assumes that everything works perfectly. If some of the poll workers don't show up and the precinct isn't running efficiently, or if a machine breaks, or if there are more candidates on the ballot than usual and it takes seven minutes to vote instead of six, well, now a six-hour line is an eight- or ten-hour line. As you can imagine, things like this happen all the time.

The infuriating thing is that the inputs of this equation were knowable before these machines were bought. Election administrators can test the machines by loading them with a sample ballot and testing how long it takes to vote. And election administrators know that at least 30 percent of the electorate shows up after 5:00 p.m. to vote. Finally, voting trends are pretty consistent, and any election administrator can look back at the last few presidential elections (when turnout is highest) and know approximately how many people they can expect to vote in their precinct in the next election. If they are really clever, they can even adjust their estimate based on population growth from previous years and predict how much higher turnout will be in the *next* presidential election.

The reason long lines became a problem with the widespread use of touchscreen machines in lieu of paper ballots is simply because the little voting privacy booths that people vote in with a paper ballot are incredibly inexpensive. They cost about $15 each, whereas touchscreen voting machines cost about $3,000 each.[5] So election administrators could buy many more privacy booths and move more voters through the line on Election Day at a much faster rate. And there is the simple fact that if the election administrators didn't put enough privacy booths in a precinct and the lines got too long, they could just give people a ballot and let them go vote in the hallway or in a corner or wherever they felt comfortable voting. The voting station is not a bottleneck when people are voting on paper ballots. When people are voting on touchscreen voting machines, the machines become a queuing theorist's worst nightmare, inadvertently disenfranchising tens of thousands of voters in the urban areas in which they are used.

I contend this is an urban and not a suburban or rural problem for two basic reasons. First, voting a million or more people in a city is a far greater logistical challenge for election officials than voting 10,000 people in a small town, particularly when you introduce voting technology into the equation. So urban areas require their election administration officials, typically the city or county clerk, to have the expertise of an industrial logistics engineer. In other words, Detroit's elections need to be run by someone with the same precision and expertise as the person running the Ford factory floor. Second, industrial engineers are decidedly *not* who we have running elections in America today. The clerk position is usually an elected or appointed position that is decided by whatever political calculations the city's political leadership think they need to make key constituencies happy. In other words, while many of these officials are elected, they are political appointment jobs because the local powers that be provide the money, volunteers, endorsements, and other political support needed to win. Unfortunately for the average voter, industrial logistics engineers are pretty far down the list of people in line for top political appointment jobs.

Desperate to head off the long lines, I asked Laurie, "Is there any way to reduce the length of lines?"

"No, it's just the way it is," she replied. "What you need to do is put together a program to convince people to stay in line."

"Okay. Well, what do campaigns normally do? Can I turn the polling places into huge dining halls where we could cater in good food and drinks, and make it into like a big neighborhood banquet?"

"No," Laurie cautioned me. "That's considered buying votes and is highly illegal."

"Giving someone waiting in a five-hour line a plate of lasagna is considered buying votes?" I queried.

"Yeah. You can only give out things considered de minimis in value."

"Okay," I said, racking my brain. "What's considered de minimis in value?"

"Bottles of water and cups of coffee," she answered. "Maybe granola bars, but you have to ask the attorneys about the granola bars."

"Water and coffee is just going to make people have to get out of line to piss," I complained.

"Well, usually people get thirsty and tired after an hour or two of standing in line. So that's why they usually give out water and coffee."

"Yeah, I would imagine they do," I responded. "Well maybe we can have big signs that say where the bathroom is and request people save people's spot in line."

"I haven't heard of the 'pissing thing' being a big problem," Laurie said. "I am sure people just save people's spots while they go."

"Yeah. Fuck. Okay. Well let me tell the staff they need to figure out how to buy and disseminate about 1.5 million bottles of water and cups of coffee across the urban and student areas of the state."

"Have volunteers do it," Laurie suggested. "Don't take staff off of contacting voters and volunteers for GOTV. If we don't hit our goals for turning out our base, we will lose the state. Obviously, we can't win the election without winning Michigan."

"Right," I said. "But it won't matter how many people we turn out to vote if they decide not to wait in these long lines and leave."

"Just have volunteers deal with the coffee and water," Laurie reminded me and hung up.

As I got my field staff spun up to mitigate the looming crisis that awaited them on Election Day, their efforts became increasingly creative but with diminishing returns. First, everyone started by trying to understand which precincts they should focus their efforts on. Some voting locations have multiple precincts voting and so were a bigger concern. Other voting locations have comparatively fewer votes and would likely not have a line on Election Day. Then there was the logistical nightmare of trying to figure out how to get enough coffee and water to 50 locations or so in a city with 500 voters at each location. This challenge was compounded by the fact that the 100 field organizers were in their 20s and even less experienced than me. On top of that, as Laurie said, their primary job had been reminding voters to vote. Each field organizer had upward of 350 volunteers and paid canvassers they were trying to organize to contact more than 1 million voters multiple times before Election Day. *That* is also a logistical nightmare. Adding the delivery of coffee and water bottles to thousands of voters waiting in line would drive anyone to hysteria. In most cases, it did. One of the young staffers was so sleep deprived from working 20-hour days, she crashed her truck into a ditch while driving to work one morning. After the crash, her main concern was not her health or the fate of the truck but getting the office open on time so some volunteers could come in and do GOTV calls to senior citizens that morning. Senator John Edwards called her to thank her for her dedication and told her to take the day off. She lied and said she would and then got right back to work.

In that state of hysteria my team came up with a lot of crazy ideas to keep voters in line. For the reasons outlined earlier, we were most worried about low-income voters, especially minorities and students, leaving the lines, so we focused a lot of our effort there. Some staff went the civil rights angle and had people with bullhorns telling the voters, "Bush wants you to get out of line." In some places they would have African American clergy go shake hands in the crowd and tell them to "have faith" and remind voters that it was important to vote. We were so concerned about the lines that when Al Sharpton was in Detroit on Election Day, we told him we preferred to have him go encourage people to stay in line where wait times were exceedingly long instead of going on TV to remind people to vote.

On the other hand, others looked at the potential crowds around the polling places as an opportunity to create a party atmosphere to interest passersby in voting. In some places, they got a local musician to play music for people as they waited. In college towns, they recruited bands to play at some of the bigger polling places. In other places, they got jugglers and other random buskers; however, my favorite idea was to send vans with loudspeakers to the big polling places after 5:00 p.m. to play dance music for the voters standing in line.

When the polls came to a close on Election Day, we had won the state by 3.5 percent.[6] Later studies showed that the impact of long lines due to inefficient election administration in the Democratic base area of Columbus, Ohio, alone resulted in over 5,000 to 15,000 people not voting.[7] It wasn't enough votes to win Kerry the state and thus the election, but that was just in Columbus. We don't know how many votes it cost him in Toledo, Cleveland, Cincinnati, Akron, and so on. Also, this doesn't factor in the almost 30,000 provisional ballots that were cast but not counted.[8] Many of these provisional ballots were cast because someone got in the wrong line and waited for hours,

• 3 •

Mandela-Level Turnout

In May 2008, I was named Barack Obama's national deputy field director.

From day one, I was on a personal mission to eradicate the long line problem. Not only did almost every battleground state in 2004 have historically massive long line problems in urban and student areas, but also we were expecting what campaign insiders were referring to as "Mandela-level turnout," a term referring to the astronomically high turnout in 1994, with the South African black population, to elect Nelson Mandela, the first black president of South Africa. By this point I had read an academic study from Ohio State University about 20,000 votes being lost in Columbus, Ohio, in 2004, due to long lines,[1] and I had this nightmare vision of what the lines in Detroit would be like this time around. Replacing the Northeastern, white John Kerry on the top of the ticket with the Midwestern, black Barack Obama would turn the several-hour-long lines from 2004 into a city-wide sea of people desperately clamoring to vote for the first African American president. Besides, there was ample evidence from the 2008 cycle that turnout would be high, especially in African American and student areas. If Obama's enormous rallies were any indication, voter turnout on Election Day was going to be huge. In St. Louis alone, he had more than 100,000 people show up to a rally.[2]

So on my first week on the job, I asked my boss, Jon Carson, the national field director, if I could be a part of the voter protection operation. This was kind of an odd request because the field department traditionally has not been involved in voter protection. That part of a campaign is normally left to a group of attorneys steeped in election law. They are generally focused on dealing with such issues as ballot design (à la Bush v. Gore in 2000), voter ID laws, and other civil rights infractions. The attorneys had been involved in the long line issue in 2004, not in an effort to reduce the long lines, but rather to remind people of their right to stay in line if they were there before the polls closed. I remember driving around Detroit on Election Day in 2004, dumbfounded at the impotence of our voter protection legal team. The attorneys were wearing these ridiculous yellow crossing-guard-type vests the campaign had bought them and handing out flyers telling people about their right to wait in line.[3] They were losing their minds grappling with the length of the lines and their total inability to do anything about it. Meanwhile, thousands of people were leaving without voting.

Standing in the campaign HQ in Chicago in 2008, I was recounting this story to Jon and telling him what my organizers were doing to keep people in line (jugglers, ministers, and sound trucks from the last chapter). Jon worked on the 2000 campaign but went into the Peace Corps after that and was in Honduras[4] during the 2004 election, so he didn't see the mess in person. But he wasn't surprised by what I was telling him: His native Wisconsin is notorious for catastrophically long lines in student and urban areas. I explained that the long line issue isn't a legal issue. It's a logistics issue. The lines form because someone in the local election department didn't correctly calculate the assets they needed (staff, voting booths, check-in stations, etc.) to administer the election.

Jon got it immediately. His rural Wisconsin drawl and mannerisms belie a steely resolve and an engineer's cold and calculating eye to every problem. Jon is one of the few political operatives I know who has a master's degree in engineering. By framing the problem as a logistics problem instead of a legal one, I could see the gears and equations start spinning in his head. He was clearly calculating how the long lines that had dogged him in Milwaukee and Madison, and other cities and state elections for years could be fixed with simple tweaks to how the elections were administered.

"Oh, shit!" Jon said. "You are totally right. This is a logistics problem. The attorneys are having a meeting tomorrow. You should go. Let me run some traps with Bauer. Have fun explaining this to the attorneys."

Bob Bauer was general counsel on the campaign, so all things legal, for example, voter protection, were run through him. At first the attorneys didn't have a problem with me being around; but as I began to turn every meeting into a conversation about long lines, everyone became annoyed. After all, the attorneys could not litigate a claim that there might be long lines. So why was I wasting everyone's time with this issue?

My retort was always, "Exactly! This is a logistics problem with a political solution. We need the political leaders in these cities to tell the election clerk they are holding him personally accountable to ensure the election runs smoothly and people don't have to wait in line for more than fifteen minutes or vote provisional ballot."

Initially, the fifteen-minute wait time goal was met with not only skepticism, but also anger. "How the hell are we supposed to do that?" was a pretty common question from the voter protection attorneys.

At first, my honest answer to that question was, "Well, fuck, I don't know."

So I turned to my field team to figure out an answer to that question. The great thing about the Obama campaign in 2008, was that you could find someone from basically every walk of life willing to volunteer their time to help elect Obama. Astronauts, brain surgeons, biologists, we had them all. I went to our tech guys and said, "I need to build a model of how to estimate voting wait times based on turnout and voting place assets (machines, staff, check in times, etc). Who can help with that?"

Within a few days I had multiple people from the data and analytics teams assigned to help me build the model, as well as volunteer statisticians, industrial engineers, and economists. I even had a Harvard MBA grad, Carrol Chang, who flew out from New York on her own dime to live in Chicago for the next five months and work on this project sixteen hours a day, seven days a week, *for free*.

Once I found the expertise to build the model, I needed the inputs to build it. The analytics team had been working on turnout projections for weeks at this point. So we had enough to start with for the turnout inputs. But we didn't have any data on what types of machines, how many staff, and what procedures each city, much less each precinct, had. So with the turnout numbers, we knew how much volume each precinct would intake. We just didn't know how many assets each precinct was going to have to process that volume.

This is again where being on the field side of the Obama campaign helped. We had statistical experts running models at HQ, but we also had more than 6,000 field staff deployed in the battleground states to put our plans into practice. My first step was to have the field team in each state send a field organizer to each election clerk's office and ask what assets they were planning to deploy to each precinct. It was a survey of voting practices and asset deployment at the most local level of government across about 20 percent of the country that was completed in just a few weeks—probably the largest third-party survey of its kind in the shortest amount of time in American history.

What we found was horrifying. In most cases, the clerks had no idea how many people and machines they were deploying to each precinct. Often the number was "about the same as last time [2004]." While our people working on the Obama campaign were obsessed, thinking daily about what "Mandela-like turnout" might mean for logistics, these clerks appeared to have given the problem little thought. Their lack of preparation was evident in other terrifying ways as well.

Organizers would come back from the clerks' offices dumbfounded and report seeing boxes upon boxes of completed voter registration forms precariously sitting around waiting to be entered into the official voter registration database. Oftentimes the boxes were overflowing or had tipped over and were spilling out onto the floor. Report after report of huge voter registration backlogs streamed into the Chicago HQ. Evidence mounted indicating this was endemic across the urban and student areas of the battleground states. Eventually we realized that we hadn't thought of this problem: Mandela-level voter registration must necessarily precede Mandela-level turnout. That is, to have record turnout, a record number of citizens would register to vote.

The surge in voter registration shouldn't have been a shock to us. It followed patterns mentioned earlier in the chapter of organic growth of voters inspired to take part in the election. Additionally, the campaign had hundreds of thousands of staff and volunteers hitting the streets, registering people to vote. Voter registration was so central to the campaign that the campaign manager, David Plouffe, spoke regularly about our commitment to not accept the battleground state map as it was presented to us. If we believed we could reasonably register enough people to vote in a red state to turn it blue, then we would do so.

What was shocking was the clerks' total lack of preparation for this overwhelming voter registration. Our field organizers didn't need a data expert from HQ to do the back-of-the-envelope math that if there are two temporary staff entering 200 forms a day and we have two months (about forty working days) until they stop adding voters to the voter registration database, then they can add about 8,000 people to the official database before the election. If we have had 10,000 voters fill out registration forms to vote, then we will have 2,000 people who don't get entered into the registration database before the election. The field staff were killing themselves to register people to vote. Standing at bus stops in 100-degree weather in August. Going to dozens of churches on Sunday. Trudging through one student dorm after another. Losing 2,000 votes in one city could cost us a battleground state and, potentially, the presidency. Couple this information with the fact that we came to the full realization of this problem at about the same time as the GOP convention, in early September 2008. Just as John McCain was getting a bounce in the polls and Sarah Palin breathed new life into his campaign, at HQ we were beginning to envision a dystopian future where hundreds of thousands of people can't vote on Election Day, and we lose the election because a handful of clerks can't figure out how to get their data entry done.

This dire but unexpected information collected by our field team about the registration backlog gave me something I also desperately needed to make any progress fixing the long line issue: credibility. For weeks I had been insisting that the inability of clerks to manage the deluge of voters that was coming on Election Day could cost us the election. Finally, I was able to demonstrate evidence from this election cycle, not just 2004, that this was a real threat. Furthermore, this isn't something that just works itself out every cycle. In fact, there are well-documented cases of these voter registration forms not getting entered before Election Day and tens of thousands of voters being disenfranchised.

An egregious example of this occurred in 2016 in Pennsylvania, where a study found that more than 57,000 voter registrations were processed late, meaning that thousands of voters did not receive a voter registration card in the mail with information on their polling place and registration status, potentially violating state law.[5] In addition, there were many instances of voter registration applications being misprocessed or *even lost* by election officials. As many as one in ten voters' names in Philadelphia could not be found on the official voter roll on Election Day in 2016.[6]

The tool Carrol Chang and the analytics team built to estimate turnout was projecting 13-, 18-, and 20-hour-long lines. Generally, the lines weren't more than five to six hours long on Election Day. Now people were taking me seriously when I explained that these predicted wait times weren't wrong because the tool was wrong, but rather because people wouldn't wait more than four or five hours, so we had eight to fifteen hours' worth of people who would be "walk-offs." Walk-offs are people who want to vote but won't get in line or will eventually "walk off" the line without voting.

The question remained: "Well, what the hell do we do about it?"

I began having the field team orchestrate a quiet but massive public engagement effort to have all manner of clergy, elected officials, political donors, and liberal interest group leaders confront their local election officials, most of whom were Democrats, and explain just how massive of a problem these long lines would be and that if the election administrator in question didn't fix it, the local leaders of his city were going to hold him accountable. I don't want to let Republican officials off the hook too much here. GOP activities around voter disenfranchisement are well documented, egregious, and frankly un-American. But that's not what this book is about. Further, what is also true is that in many Democratic strongholds like student towns and bigger cities, elections are administered poorly. It is Democrats who are in charge of administering elections in those cities and often are wholly or in part the ones to blame for those disenfranchised. We knew this well and we were not going to let anyone off the hook just because he was a Democrat. Fortunately, there was no place for these Democratic election officials to hide from us. In a world where we all live in our own bubble, the hardest thing to escape is not when the other tribe's bubble is after you, but when your own tribe's bubble is after you. For Democratic election administration officials, we owned their bubble. We were relentless and we never went away.

Moreover, we worked closely with Bob Bauer's two deputies, Kendall Burman and Mala Adiga. They ran the legal side of voter protection work for the campaign. So when my twenty-five-year-old field staffer was being blown off by a clerk who didn't want to explain his asset deployment plan for Election Day, she had an experienced lawyer backing her up.

With the operation beginning to fire on all cylinders, the scope and impact of the election administrators' lack of organization came into focus. For example, in Indiana, we were concerned about the potential for incredibly long lines because the state hadn't been a battleground state since the 1960s, when the state voted for Lyndon B. Johnson.[7] So on top of the fact that we were concerned the election administrators in general weren't ready for "Mandela-level turnout," we were even more concerned that those in Indiana weren't even ready for normal presidential battleground state level turnout. We tried to attack the line problem from multiple angles. The first step in understanding the unique problems in a county or state was to have our data guys help us figure out where the choke points were, like the check-in station or the voting booth or sometimes even a lack of parking spaces, and address those issues in turn. Another way to deal with the line issue was to get more people to vote before the election, thus taking them out of line on Election Day. It turns out that early vote operations were as poorly managed by some election administrators as data entry of the voter registration forms and the lines on Election Day. The most obvious example of this was in Indianapolis, where multiple early vote stations were open on the outskirts of the city on nights and weekends during times convenient for most people to take advantage of them. Conversely, there was only one early vote station in downtown Indianapolis, and it was only open from 10 a.m. to 4 p.m., Monday through Thursday.

Most urban areas in the United States in 2008 still followed the traditional demographic pattern of poor minorities, in Indianapolis's case, African Americans, living closest to the downtown and more affluent white people living further away from the city center. Most Midwestern cities, save Chicago, had not started following the pattern of the coastal cities like New York and San Francisco, where the most affluent white people live in the city center and the poor have moved to the outskirts of the city or suburbs.

Essentially, rich white people on the outskirts of Indianapolis had multiple places to vote at times convenient for them, and poor black people downtown had one place to vote that was only open during the *most* inconvenient time of the day. We pleaded with the Marion County clerk, a Democrat, and her staff over and over again to fix this. We pleaded with her to open more early-vote sites downtown and have them open on nights and weekends, just like they were in the affluent white areas. We didn't want any more access for African Americans than white people had, just the same level of access. We made our case to the clerk through visits from our newly integrated voter protection staff, local Democratic leaders, and even African American clergy making a moral appeal. We spent months trying to explain how unfair this was, and how much of a problem it could cause in terms of the subsequent long lines on Election Day and how tens of thousands could be disenfranchised if they couldn't wait in line, not to mention how infuriated her voters would be if they couldn't vote for the first African American president.

Over and over again, we got some variation of the same response we had gotten from dozens of other clerks offices throughout the country: "We don't have the resources to expand," or "We have always done it this way," or "Who are you outsiders to come in and tell us how to run elections here?" Then there's my favorite: "Voting is people's civic duty; they should be proud to wait in line to vote."

After months of pleading, we decided we were getting nowhere and needed a new strategy. Eventually we settled on two basic principles: First, getting past the argument that "We have always done it this way," was going to be harder than we thought. This was compounded by the fact that these election administrators weren't racist. They just didn't believe that their practices could be disproportionately disenfranchising minorities. Second, I argued strenuously that we were going too easy on the election administrators. We had a strong body of evidence that Election Day was going to be a complete disaster: the recent history of catastrophically long lines in 2004, hundreds of thousands of voter registration forms that would not be entered before Election Day, and now early vote sites that ensured low-income minorities or students did not have the same access to early voting opportunities as affluent white people. Their incompetence in planning and logistics would disenfranchise hundreds of thousands of people, most of them Democrats and most of those Democrats poor minorities and students.

Finally I said to my team, "Look. I know the election administrators are complaining that we are being assholes and the voter protection people in each state are getting grief from them; however, this isn't like some recalcitrant Democratic congressman who won't agree to cohost a rally for Obama." I continued,

> We have now explained to all of them how their actions will disenfranchise thousands of voters in their city. So these people are knowingly choosing to stop people from voting. So fuck 'em. The gloves need to come off. We need to use everything we can legally and ethically use to force them to ensure people can vote.

So in Indianapolis, for example, we settled on a two-part plan: We had our graphics people put together an infographic depicting the stark differences between where the early voting was for affluent white people and where it was for poor black people. It basically looked like census maps showing different-colored blocks where black and white people lived and early vote sites with hours of operation plotted on the map. Our graphics team did a great job. It was incredibly simple yet powerful. The red indicated African American voters, and yellow indicated white voters. The only early vote site anywhere near the red areas was the aforementioned site that was only open during the workday Monday through Thursday. On the other hand, early voting sites with evening and weekend hours dotted the yellow areas. Our team took this graphic in to the clerk's office and explained that we would leak it to the press—all of whom were dying for scandalous stories about voter disenfranchisement in the battleground states. Within hours of it leaking, we predicted, the graphic would be a scandal running on national television.

For the second part of our plan, we went to several of the key ministers in the African American community and asked for them to stand with us in convincing the clerk to ensure their parishioners had equal access to early voting. Fortunately, the ministers were more than happy to help with this effort and said they had years of voting access infractions they were frustrated about.

Then we took it to another level. We decided to stop wasting our time with the clerk. Instead we just told her that Senator Evan Bayh, several African American ministers, and other activists would be leading a march to vote early over the weekend. But we would be marching not to the early vote sites in the white areas that were open over the weekend, we would be showing up at the one early voting site in the African American part of town. As stated above, that site was closed over the weekend. We said we would be bringing our handy map of the city showing the disparity between access to early voting for African Americans and whites. Oh! And would also be bringing every TV station, newspaper, radio station, and random blogger in town. The event could wind up being great publicity for her, showing how she was making voting convenient for everyone and access equitable. Or, it could turn into a protest outside of her office, led by the Senator and many ministers and civil rights leaders denouncing her on camera.

The clerk went apoplectic.

She fumed that the building wasn't open then and she had no control over when it was open or not. We said we didn't care. If she wanted any support from these ministers and political leaders in the next election, she would fix the problem. We were done asking her to fix it, we just told her we were coming. She knew the press would not buy that there was any acceptable reason the offices in the white areas were open while those in the African American neighborhoods were closed. She would be the face of this injustice, especially if there were thousands of people protesting outside her office.

After months of trying to get through to the recalcitrant election administrator, the situation was fixed in a few days. The downtown location added dozens of voting machines, opened on nights and weekends, and added more locations downtown.

Another favorite example of how our newly congenial field/voter protection team worked together was in Virginia. One of the choke points that invariably presented itself in Virginia was the use of touchscreen voting machines or lack of voting privacy booths. After racking our brains for weeks trying to figure out how we could get around the bottleneck of a fixed number of touchscreen voting machines or privacy booths in each precinct, we eventually came up with an answer that was both obvious and simple. We would just recommend that once the line is more than thirty minutes long, the precinct workers should begin adding more privacy booths and voting people on non-provisional, real ballots that would be counted that day, just like absentee ballots are. Essentially, the paper ballots and booths would become the release valve for the backlog caused by the touchscreens, because if the lines are really bad, you can have as many voting stations as you do ballots. Basically, the precinct worker can say, "If you don't want to wait in line, just take a paper ballot and go vote in a privacy booth and hand your ballot to the precinct worker on the way out." Obviously, privacy is critically important, but privacy booths are incredibly cheap to buy. It was easy, simple, legally compliant, and the logistics were already in place. Almost all states had a provision for some paper ballots to be kept onsite in case all the machines broke or some other "disaster" occurred. So all we had to do was get the local registrar to agree to have more ballots and privacy booths at each location and to train the poll workers that a line longer than a half an hour was a "disaster," triggering the paper ballot option and increasing voting stations. Next to trying to get more people to vote early, this became our standard solution for long lines.

Despite this being an incredibly simple fix for an incredibly complex problem, we had to go to war with many of the registrars before they would agree to it. One of the best examples of the lengths we went to ensure voters didn't have to wait more than 30 minutes to vote was in Chesterfield County, Virginia. The Chesterfield registrar adamantly refused to make this simple fix despite there being a long history of voters forced to wait in several-hour-long lines on Election Day.

As usual, the registrar told us that he didn't want us telling him how to do his job. We explained that we didn't want him to do anything unreasonable, like spend a bunch of money he didn't have to buy hundreds of new voting machines. Rather, all we wanted him to do was let people vote on real, not provisional, paper ballots when the line got more than fifteen minutes long (our goal was to get the registrars to agree to thirty minutes, so we started by asking for fifteen, assuming we would have to negotiate this point with them). He repeatedly told us to go pound sand. Obviously, we were used to this by now so it didn't dissuade us. This was Virginia—there was no way we would give up.

Virginia and Colorado were ground zero for the campaign. The economy was devolving quickly into the Great Recession, as it later would come to be called. McCain had not yet commented that the "fundamentals of the economy are strong."[8] At the time, we had no idea how the economic downturn would affect the race. Everyone was cognizant of the dynamics that had played out in 2000 and 2004, when there were more than a dozen states on the battleground list, but in the end it came down to Florida and Ohio, respectively. Our "Path to 270," which is the number of electoral votes one needs to become president, ran directly through either Virginia or Colorado. Essentially, our analytics guys decided that based on demographics, polling, and so on, if we won the same states Gore and Kerry won, our best path to 270 was to win Virginia or Colorado. This clerk didn't realize it at the time, but he was dealing with a group of people who literally believed that the future of the free world might come down to hitting our vote goal in Chesterfield County, Virginia. So long lines that disenfranchised voters were not okay. In other words, we weren't going anywhere.

We went back to the drawing board, knowing that we were running out of time. We were down to the last two weeks before the election, so we began scrambling to figure out how to get this obstinate guy to come around. Colin Bishopp, who was from a local labor organization, came up with a simple yet ingenious idea. About a week before the election, he made T-shirts that read, "Thank you for waiting in line for me." Then he recruited some of our African American volunteers' kids (Chesterfield County has a large African American community) to wear the shirts on Election Day and hand out flyers to anyone waiting in lines. He took a picture of the African American kid in the T-shirt and brought that and a copy of the flyer to the registrar.

The clerk looked at the picture and began fuming. Then he looked at the flyer and freaked out. He was absolutely flabbergasted pleading with Colin along the lines of, "This is unacceptable! I can't believe you would pass something like this out!" And so on. The reason the flyer set him off was that on it was his picture and phone number. It read,

> YOU SHOULD NOT HAVE TO WAIT FOR HOURS TO VOTE! Chesterfield County Registrar . . . has ignored warnings about the expected heavy

> turnout on Election Day. Call him and tell him that you deserve the right to vote in a timely fashion: [his number was inserted here]

Below that it had the pictures of the five elected officials who had appointed him to his position, and it read, "Chesterfield County supervisors refused to take steps to correct this issue. All supervisors are up for election this year," and encouraged voters to vote against these people who appointed the registrar to his job.

Colin didn't call the clerk or e-mail him the picture and flyer. He went to the clerk's office and sat there, face-to-face with him. He showed him the copies of the picture and copy of the flyer and said,

> Sir, we've been trying to be reasonable. We've been asking you to take reasonable steps that don't cost any money to reduce these lines. All we want you to do is execute the minimum requirements of your job. All we want is to ensure that everyone legally allowed to vote in this county can vote and that people don't have to wait hours and hours to do so. If you're not going to work with us to ensure that happens, then we really are left with no choice but to let the voters hold you and the people who appoint you to your office accountable.

It was a balls-of-steel move.

And it worked. Two weeks later on Election Day, when the lines got long, voters were able to vote on real, not provisional, paper ballots and there were plenty of privacy booths. Pete Rouse, Obama's senior advisor, loved the move so much he kept the flyer on the wall in his office throughout the rest of the election.

These were just a few anecdotes out of hundreds like them. Every day we were fielding constant reports of disorganization in election administrators' offices. There were daily battles with stubborn election administrators who refused to make basic changes to fix processes that had completely broken down. There were ongoing revelations of total inability to implement basic logistics planning and resource allocation. With all this in mind, I vividly remember cramming into a "phone booth" office at Obama's Chicago HQ two weeks before Election Day to discuss voter protection issues. One of our many tech guys piped up and said,

> Have we thought at all about how easy it would be for someone to hack into one of these election offices and change the results as they are being reported? Or delete or scramble names in the voter registration database so thousands of people wouldn't be able to vote? A group like the KKK would surely be motivated to disrupt the election of the first African American president and could potentially have the capability to hack the election infrastructure.

I rolled my eyes and said,

> Jesus, are you fucking kidding me? The election is two weeks away and I am still trying to get these fucking guys [election administrators] to just do basic data entry and put enough voting booths at each polling place to handle the volume of voters. If a group of hackers attacked them, we are fucked.

So we just moved on and didn't talk about it again.

Little did we know that eight years later, these same election administrators would be our only line of defense against Russian hackers trying to disrupt our elections, far more sophisticated than anyone in the KKK.

• 4 •

Reset

In 2009, nine months into the Obama administration, I begged for a week's vacation to recharge, as I had been working 100-hour weeks since the primaries began in 2007. My vacation was about six months after then-secretary Hillary Clinton's ill-fated "Reset" meeting with Russian Foreign Minister Sergei Lavrov. In this meeting, Clinton comically presented Lavrov with a small device with a big red button, which read "reset" on it in English, but it was inaccurately translated to "overload" in Cyrillic.

The gimmick was intended to signify an attempt by the Obama administration to start over with regard to U.S.–Russian relations. In fact, concessions were made by both sides when Putin agreed to let U.S. war planes fly over Russian airspace on their way to Afghanistan, and Obama agreed to stop building a missile defense shield in Eastern Europe, a program liberals had been opposed to since the Reagan administration. This made Russia an interesting place to travel for a vacation. It was, like the United States under newly elected President Obama, a place that seemed to be undergoing massive change. My wife and I were also incredibly excited to see the historic architecture and museums there. But during my security briefing I began to get the sense that I was overly optimistic about how much change was really occurring. This was typical of young Obama staff like me—intense hope and optimism for change, and then frustration and surprise as to how difficult and slow change can be. So, I was surprised when I heard the briefer say something along the lines of, "Because of your title they will view you as if you are the equivalent of the Kremlin's representative to the KGB."

"What?"

"Well, White House liaison to the Department of Homeland Security could be construed by them as someone with a similar job to their Kremlin, KGB rep."

"Okay," I replied. "But all I do is political appointments and public engagement."

"That's not the point," was the response. "The point is that you are a target. So you and your wife should assume you will be targeted by their intelligence officers while you are in Russia."

"Oh okay. Well are we in physical danger? Should we cancel the trip?"

"No. No. They don't want to harm you. They want to compromise you in some way to get dirt on you and then have you become a source for them. So don't do anything stupid. No drugs. I am glad to hear you are going with your wife since they will often try and get to you with women." The briefer continued, "Just keep your wits about you and stay out of any sketchy situations cause they could be a set up to try and get you to do something dumb. Oh and keep your electronics on you at all times. I don't mean in your room. They will search your room. I mean on you."

"Huh. Okay. Well you don't have to worry about drugs and women. I will keep my antenna up for sketchy situations."

The rest of the briefing is classified; but I have no idea why because he didn't tell me anything I couldn't have figured out by reading the *New York Times* or *Wall Street Journal.*

As Jena and I got off the plane, we were struck by how much Russia really looked and felt like the old Hollywood movies about Soviet times. Color is not something Russians seem to gravitate toward. On top of the constant overcast weather, the place is made all the drearier by the gray concrete coloring that seems to wrap every building, coupled with the fact that almost everyone is covered from head to toe in black clothes. And since we were there in October, any greenery that might have existed was stripped down to its bark and blended in with the dank atmosphere that surrounded it.

To further complete the picture, everyone smokes nonstop, and they smoke everywhere, even in the airport, and pollution is ubiquitous. And then there is alcohol. Russia's national alcohol problem is well documented, but when it's staring you in the face in the form of a half-dozen twenty-somethings of prime working age sprawled out on the sidewalk at one o'clock in the afternoon on a Tuesday surrounded by multiple bottles of vodka and drunk out of their mind, it furthers the sense of a bleak, sparse environment. This may be why Red Square and St. Basil's Cathedral are so popular; aside from being stunning works of architecture, they are the vividly colorful exceptions to the rule.

So it seemed out of place in this melancholy environment when we walked out of the metro and two young police officers walked up to us and said in perfect English, "Are you lost?"

To be sure, we have traveled extensively. We knew to dress like the locals (all dark colors), no flamboyant Hawaiian shirts or cameras around our necks. And we weren't fumbling through some big map that would indicate we were lost. This was before everyone used Google Maps to get everywhere. Finally, we are both white, so it's not like we stuck out because of our ethnicity like we would in China or elsewhere.

Jena gave me a look like, "What the hell is this?"

I remembered that the security guy said we were not in physical danger, so I decided to engage. I was pretty sure we knew where we were going, but figured I would ask them if we were going in the right direction. One thing led to another, and these guys were giving us a ride to our hotel and cracking jokes on the way.

I came to realize later that this is pretty standard stuff. Russian intel doctrine is such that they want you to know that they know what you are up to, no matter how random it is. So instead of initially engaging us as we got off the plane or in our hotel, they picked a seemingly less predictable place to interact with us, as we were getting off a random metro stop.

The two officers dropped us off at our hotel without incident, and we checked in. It is also possible that these were just two nice guys or maybe we stuck out for some reason I am not remembering. Maybe our map was easier to see than I thought. Maybe they heard us speaking English to one another and just wanted to practice their English. Really there are a bunch of reasons besides surveillance these guys could have come up to us.

Through some colleagues at DHS in the embassy, Jena and I were connected to a few guys from the State Department at the embassy in Moscow. One of them was Brendan Kyle Hatcher, a diplomat who had just been through a harrowing series of events with Russian intelligence. He was approached by Russian spies who said they had a sex tape of him with a prostitute at a hotel in Moscow.[1] He said that was impossible because he had never slept with a prostitute in Moscow or otherwise and was happily married. The spies then produced a tape that showed him walking down the street in Moscow and another scene of him walking into a hotel. The next scene shows a woman walking into the same hotel. The following scene shows two people who vaguely resemble Kyle and the woman more than fooling around in the hotel room.

Over beers Kyle told us the spies demanded he become a mole for them. He told the spies that while the tape proved he had walked down the street in Russia and had at some point walked into the foyer of the hotel, it proved nothing else, as the two people in the hotel room could have been anyone. The spies told him that if he didn't comply, they would release the tape. He effectively told them to go pound sand and then followed State Department protocol by alerting his superiors to the incident.

Kyle was infuriated less at the Russian spies, as they were just doing their job, and more so at the American media. After he refused the Russians, an

American news outlet, ABC, ran the story,[2] instead of refusing to play into Russian ploys to smear a U.S. diplomat into compliance. This was not only un-American, but also bad journalism, allowing the outlet to be used as Wikileaks would be a decade later. A witted or unwitted pawn, but nonetheless a pawn in a Russian intelligence scheme against the United States.

This type of coercion, while very clumsy, was relatively standard stuff for intelligence agencies to carry out against its nation's adversaries. It stood in stark contrast to the image Clinton and Lavrov were displaying for the world at the Reset event. This was not the type of stuff allied countries, which we supposedly now were, do to one another. Nearly all nations spy on one another. That being said, such coercion techniques as threatening to ruin a person's marriage and end their career via a fake sex tape is considered the type of thing adversaries do to one another, not allies.

We felt horrible for Kyle; he was clearly under a lot of stress because of the ordeal. However, he was a real diplomat and a pro, someone in whom the American taxpayer can and should be proud, and to his credit he took it in stride. Hopefully the beers and laughing about it for a while helped.

While we were in Russia, I became acquainted with the friend of a friend working for a legally registered international NGO in Moscow that works to build and strengthen civil society and support democratic processes.

This acquaintance's (let's call him John at his request for anonymity out of concern for friends and colleagues in Russia) experiences in Russia reflected relations that did not at all resemble the backslapping days of President Clinton and Yeltsin. John worked with local civic activists, researchers, and governmental officials on a variety of issues that touched on good governance, election processes, and perceptions of society.

John was a welcoming and hospitable guy, who went out of his way to show us around Moscow. He made sure we got off the tourist circuit and experienced some authentic local food, and even took us to a hilarious hockey game replete with American style cheerleaders who were mimicking the outfits correctly, but didn't really get the cheering part right. Over *pivo* (beer) and great Eastern European meals, John regaled us with tales of constant surveillance and harassment in Russia. He described trips to the regions where skinheaded thugs, or black sedans openly tracked their movements. The team's e-mail systems were hacked, and new partner organizations were dissuaded from future cooperation after visits.

The harassment did not always stop at intimidation, but could escalate to violence. Incidents of physical attack were never clearly attributable to the government, and could individually be explained away as typical urban crime, but as any Russian activist could attest, the pattern of behavior is easily identifiable. In a six- to eight-month period, one member of the team was attacked and beaten three times, another was beaten coming out of a metro, and a colleague at a different organization was attacked on the street his first night in Moscow. Paraphrasing John: "Sure, Moscow has street crime, like any city, but the central city is fairly safe and this kind of repeated violence doesn't happen and isn't a coincidence."

Fair enough. This didn't sound like the activity of a country that wanted to "reset" and open up to the West. It sounded like a thuggish authoritarian regime that was oppressing its people to stifle progress.

A few days later, we met up with the lead of the DHS mission at the U.S. Embassy–Moscow. While I was on vacation, I figured that offering to take the top DHS guy in Russia out to dinner would be a good gesture from the Obama administration. In addition, these guys don't get much attention from D.C. headquarters, especially not the White House or the Secretary of Homeland Security's Office. I got connected to him through Chuck Marino, who sat across the hall from me in the secretary's suite of offices and was the Secret Service representative to the secretary. I thought we would end up talking about what the new Obama administration policies would mean for him and anything else related to Secretary Janet Napolitano's specific DHS policy plans that would affect him.

As usual, I was totally wrong. He didn't want to talk about any of those things. This man, whom I'll call "Jim" (not using his real name here to protect his privacy) is responsible for getting counterfeit U.S. dollars out of circulation in Russia. As soon as we sat down and exchanged pleasantries, he began telling similar stories of surveillance and harassment. In the first 30 minutes, he had identified at least three people in the restaurant he was pretty sure were there to spy on him or me or both of us. Like most restaurants in Moscow, the establishment was dimly lit. Tables were tucked off in corners. So it seemed everyone was either trying to keep their conversations hushed or hiding in an effort to spy on someone. This was, of course, exacerbated by the power of suggestion. When a six-foot-four Secret Service agent in Moscow says you are being watched in a dimly lit restaurant, you assume everyone is a spy. Every glance from the waitress to see if your drink needs a refill feels like an attempt to read what you are typing on your phone. Every patron sitting alone at a nearby table or person walking into or out of the bathroom at the same time as you seems like a spy keeping tabs on your every move.

Jim's mandate to get counterfeit U.S. dollars off the street wasn't the surprising part of his job. That is a core function of the Secret Service. Protecting the president and other high-ranking officials is only a small part of what they do. Allies would normally want help in finding and taking down local counterfeiting operations. He had Russian counterparts who, as far as he knew, were genuine partners against these counterfeiting operations. But they were a small minority with whom he dealt. Still, the constant surveillance and harassment was incongruous with how allies normally treat allies.

This is why another American living in Russia told me he never understood what "freedom" really felt like until he did a short holiday in Western Europe after working for a year in Russia. He said that once he stepped off the plane in Germany, he had this sense of liberation in that he was confident no one was watching him. No one in the German government really cared what he was doing. He was just another tourist sightseeing. This paranoia became a pall that was clearly hanging over every American I met there. But I guess you are only paranoid if they are *not* constantly spying on you. These folks weren't paranoid, they were traumatized, so much so that, sadly, Jim told me his wife and kids were moving back to the United States because they couldn't stand the unrelenting surveillance. As a true patriot, he was committed to his mission there. His assignment wasn't over. It pained Jena and me to witness this hulk of a man crumbling inside at the thought of being away from his family for the next year. At the same time he was relieved they would be out from under the shadow of surveillance in Moscow.

One Sunday night, Jena and I decided we wanted to go out and experience some of the nightlife in Russia. We did not plan well, as Sunday is not the best night to find a party. But we didn't seem to be alone in the search that Sunday. We asked our hotel where we should go to find a few bars and clubs to check out. They sent us to a strip that had a bunch of bars and nightclubs. Only one was open. This made our decision-making process easy. We headed in for a drink. The street was basically shut down, and few streetlights and storefronts were keeping the street lit. Like many big urban areas, once you get off the main drag and to more residential areas, it seems eerily quiet and secluded. But this was supposed to be an entertainment district. We were a little put off that the entire block seemed to be shut down except this one bar. That being said, it was Sunday night, so what did we expect?

I went up to the bar to get us a couple beers. Before I returned two guys were already talking to Jena. This didn't strike me as unusual. Jena is an attractive woman. She can't sit at a bar by herself for more than a few minutes without a guy coming over to strike up a conversation with her. I never take offense at this. She is clearly wearing a wedding ring, but whatever. Maybe the scores of guys I have seen come up to her throughout the years since we got married failed to notice her ring.

I walked up with the beers and said, "Zdravstvuyte," which means "hello."

We began talking and quickly made fast friends. They asked what we were doing in Russia. We said we were on holiday. I didn't offer that I was with Homeland Security, but I did mention that I had worked for Obama.

Soon they were buying shots of vodka as is customary in Russia. We were just trying to keep up. After a couple rounds, I got up to go to the bathroom. A couple minutes later the most talkative of the two guys came in after me. Instead of walking straight for the toilet, he walked up to me and pulled out a 3-inch-thick wad of $100 in bills.

My antenna was already up because of the dark street that was supposedly an entertainment district and the fact that two guys came over to us immediately upon entering the bar, not to mention they had been quite generous and even pushy with their interest in buying us vodka shots; but again, that is not totally out of character for locals in any country when meeting foreign tourists. If you travel a lot, you are used to folks you meet in bars shoving drink after drink in your face of their local libation of choice. What is totally out of left field is for them to follow you into the bathroom and show you a wad of $100 bills.

This seemed exactly like the situation my DHS security briefer had referred to. It was totally sketchy. I was less than alert, as they had been pumping us full of liquor, and this guy had cornered me in a bathroom and was standing there with a lot of money in his hand in front of a mirror. I began thinking they probably had multiple cameras on the other side of the mirror. As my mind was racing, I was thinking, "Whatever you do, don't touch the money." I figured they just wanted a picture of me handling the money and then they would use it to blackmail me like they tried to do to Kyle with the fake sex tape.

I put my hands in my pockets. Instead of offering me a bribe, which is what I was expecting him to do, he cleverly asked me what to make of a red stamp on the bills. This seemed like a much more clever way to get me to handle the money. This way he could trick me into handling the money by asking me to inspect this stamp and not have to bet on me taking a far more extraordinary step of accepting a bribe. Similar to their strategy with the case with Kyle, the Russian intel services had decided they didn't actually need to do the more difficult work of finding and entrapping people doing treasonous, illegal, or immoral activities. They could just make it *look* like the person had done something untoward and that could be used against them instead.

I kept my hands in my pockets and told him I was sorry but I had no idea what the stamp indicated. He was standing between me and the bathroom door, so it was hard to just walk out in a way that wasn't awkward and wouldn't risk escalating the situation or getting so close to him that he could make it look like I was reaching for the money, when, in fact, I was just trying to get out of the bathroom. I just stood there. He kept pressing me to help him decipher the meaning of the stamp. His story was that he had been paid in cash by an American businessman that he was working with, but now he was worried the guy was paying him in counterfeit dollars and that he was risking arrest by depositing the money into his bank account.

After he gave up trying to get me to touch the money, we went back out to the table we were at and tried to get Jena to touch the money. She wisely kept her hands on her beer or her lap, and he gave up for the moment. After a few minutes we decided enough was enough. We said it was time for bed and got up to leave. They offered us a ride, and we initially demurred; but after waiting for a while on the empty street and deciding that since my security briefing advised that we were not in physical danger, we agreed to take the ride.

Once in the car it started up again. But this time on top of pulling out the money again, he offered us drugs. "Coke, weed, ecstasy, whatever you want. C'mon man you wanna party?!?" We did not.

He kept pushing and told us that his far quieter friend who was driving had just gotten out of prison and could get us whatever drugs we wanted at a very reasonable price. Again, I began looking around the car to try and spot cameras or microphones. While not seeing anything, I assumed that, having failed to get a picture of us handling the likely counterfeit dollars, this was his last-ditch effort to get us on tape doing something illegal—in this case, buying or consuming drugs. Of course we refused, and eventually we wound up back at the hotel without further incident. He also said if we wanted to contact him we could just ask the front desk and they would know where to find him. Again, this seemed like a Russian intel tactic to remind us that he was already working with the hotel owner and we were being watched regardless.

Once I got back to the United States, I reported everything that seemed of issue with my security officer at DHS—not the least of which was that *all* of our electronics, which were with us the entire time, seemed to have been hacked. Our phones, tablets, all of it began showing a notice saying, "A media card has been inserted into your device." After reporting this to the DHS official, along with the information that we had our devices with us, in my backpack, at all times, he said we needed to deliver our electronics to them for forensics purposes. They wanted to study the malware the Russians had put on our devices. I asked how they could have done this when we had our equipment on us for the entirety of the trip. I was still pretty naive about how easy it is to hack into electronics, especially anything that touches the internet or touches parts that touch parts that touch the internet. He said that while we may feel like no one had physical access to our devices, we weren't giving the Russians credit for how good they are at their spycraft. For example, the incident in the bar with the two guys and the money could have been a diversion tactic for the quiet guy to put malware on our devices while we were paying attention to the guy who seemed to be trying to entrap us. Also, any of the devices that had a connection to the internet, even if we didn't have it turned on, could be hacked by Russian agents using Russian cell towers.[3]

While this type of activity happens all the time between adversaries, one thing that didn't seem to be happening was any sort of "reset." The one thing I came away from the Russia trip with above all else was that the Russians are a tough and proud people who were not going to just cede global leadership to the United States. And they didn't seem to be acting at all like allies.

• 5 •

Same Old, Same Old or Something New

So far we have learned that elections in the United States are not the country's most efficiently run activity. Further we have also learned that Russia behaves less as friend than foe. However, the subsequent question we must answer is, how do these two seemingly unrelated facts manifest in a way that affects your life? Many people would answer that question by pointing a finger at so called "fake news" disseminated by the likes of Facebook, Twitter, and YouTube. But Facebook, Twitter, and YouTube do not administer elections. The purpose of this book is not to address the myriad ways Russians used social media in the 2016 or 2018 elections to affect the electorate. There are several other books that delve into Russian use of social media in the campaign from fake news to the bots that amplified it; however, to understand the broader influence of the Russians in 2016 and fully contextualize the scope of their much more insidious attacks on our election administration infrastructure, it is important to briefly contextualize the role of social media in the 2016 Russian attacks. To that end, here I give a brief overview of how the Russians used social media as part of the continuum of their broader influence campaign to impact the election. I will also go deeper into how they attacked our election administration system itself.

But first, a basic question needs to be answered. Did any of the Russian activity actually help elect Donald Trump? Were voters persuaded by Russian fake news? Were votes deleted, added, or changed by Russian hackers? Most importantly, what are the types of attacks we should spend our resources to guard against in the future?

My answer is that the Russian fake news and bots on social media probably didn't affect the outcome of the election. Also, we do not and will not ever know if the attacks on the election administration infrastructure in 2016 changed, deleted, or added votes. We should spend our resources to fight both the fake news and hacks on election infrastructure. However, attacks on our election infrastructure are potentially far more damaging to our democracy. Not trusting what's in your Facebook News Feed is one thing. Not trusting that your vote will be counted as it was cast, is something categorically graver.

SOCIAL MEDIA

Early analysis of the 2016 Russian attacks made clear social media played a significant role in the disinformation campaign employed by Russian intelligence. They planted false information to influence voters' opinions of candidates, as well as instilled distrust in American institutions to fairly administer the election. A report released by U.S. intelligence agencies made it clear that this was a primary objective of the attacks, stating that Putin's goals were to "undermine public faith in the U.S. democratic process, denigrate Secretary Clinton, and harm her electability and potential presidency."[1] News outlets reported that the "NSA intercepted Russian officials congratulating themselves after Trump won the election."[2] So it would appear that the campaign was successful, at least in the eyes of the Russian government.

For months, more details were provided about the extent of the social media influence campaign. According to congressional testimony from Facebook, Twitter, and Google, as many as 126 million Facebook users may have seen Russia-sponsored content; 2,762 Twitter accounts were controlled by Russians, and more than 36,000 Russian bots tweeted 1.4 million times during the election cycle; and as many as 1,108 videos lasting 43 hours concerning the Russian disinformation campaign appeared on YouTube, as presented by Google. Google also found $4,700 worth of Russian search-and-display ads on its search platform.[3] The tech companies also stressed difficulties they faced in keeping up with the threat of foreign intelligence. In response, Facebook launched a tool to allow users to see if they interacted with Russia-sponsored content.[4]

To summarize, NSA's report basically posits that a lot of people saw Russian content about the election. NSA doesn't comment on whether the content had an impact. Note the careful language used by NSA in the previous paragraph. It merely lists content posting stats.[5] NSA doesn't say the posts actually targeted swing voters or GOTVed GOP voters for Trump. Furthermore, NSA doesn't have the legal authority or electoral expertise to research the necessary data sets to assert an informed opinion on this topic. Answering that question would require an enormous amount of research on U.S. citizens living on U.S. soil, which NSA largely cannot do (aka all the trouble they got into after the Edward Snowden hack).

But NSA's opinion is beside the point. We already have the data we need to answer the question. If swinging undecided voters or GOTVing GOP voters for Trump was Putin's goal, he failed miserably. In 2016 vote totals as a percentage of the total electorate in the three states that cost Hillary Clinton the election (Michigan, Wisconsin, and Pennsylvania) was down for *both* candidates from the previous three presidential elections.

Professor Anthony Fowler at the University of Chicago has done some incredibly insightful research that shows social media has no demonstrable effect on turnout.[6] Alternatively, some people may argue that the Russian bot–propelled fake news worked to decrease turnout among African Americans in those three states. There is no academically rigorous research indicating that while social media does not *increase* turnout, it is effective in *decreasing* turnout. So while it is possible that Russian bots depressed African American turnout in those states, there is no research showing social media has any power to decrease turnout.

Conversely, there are mountains of research, including that of Professor Fowler, showing human-to-human contact has an effect on voter behavior. Fowler's study shows such in-person contact as calling people on the phone or talking to them face-to-face at their doorstep has a significant impact on their likelihood of turning out to vote. Furthermore, in 2008 we found evidence on the Obama campaign that corroborated the aforementioned study. In fact, in 2008, through millions of human-to-human contacts with voters, polling, vendor phone surveys, and data analytics, we found that swing voters who had spoken directly to a volunteer or other representative of the campaign performed more than 10 points better than other voters to whom we had *not* spoken.

So, while there is no evidence Russian bots persuaded swing voters or GOTVed GOP voters for Trump, there is ample evidence that Clinton's less-than-robust campaign in those three states simply talked to fewer voters than did John Kerry, and Barack Obama twice.[7] Thus, fewer people voted for her in those states. The result: Despite Trump getting a smaller percentage of the electorate than his three predecessors in those three states, she still lost.

So the ongoing hysteria about fake news and bots may be misplaced. The fake news and bot activity may simply be the modern version of Soviet operatives passing out flyers at communist rallies and on college campuses in the United States during the Cold War. It is certainly something our national security agencies that monitor and combat foreign propaganda should work to stop; however, it may not really have much impact on the outcome of elections, just as the dissemination of communist propaganda to support communist parties in the 1950s had little effect on the outcome of elections then. To be crystal clear here, I am not asserting that we "know" categorically that social media did not or cannot have an impact on voter behavior with respect to if or how they vote. What I am asserting is that there is not yet rigorous, academic, peer reviewed research that has found these tactics to work in changing this type of voter behavior.

ELECTION INFRASTRUCTURE

On the other hand, what is wholly new and a far greater threat to the foundation of our democracy are attacks against the infrastructure used to *administer* an election. To my knowledge, Soviet operatives never stuffed ballot boxes in the United States during the Cold War; however, deleting voters from voter registration databases so they are not given a ballot when they show up at a polling place can have the same effect as stuffing a ballot box. That is, taking would-be votes away from one candidate has the same effect as stuffing extra votes in the ballot box of his opponent. Additionally, an NSA whistle blower leaked that Russians successfully hacked at least one voting-machine company.[8] We know the Russians at least researched hacking the vote-tallying machines as well. That would allow them to change, delete, or otherwise alter votes and therefore vote tallies.

But what Putin is likely trying to do here is not elect one candidate over another, but rather undermine public trust in the election process itself. In fact, it seems this is what Putin was planning to do in 2016, and Trump's victory may have been an unintended consequence. It is highly unlikely that Russian intelligence officers are any better at predicting the outcome of U.S. elections than the U.S. data scientists and campaign professionals who spend their entire careers and billions of dollars reading the tea leaves of the American electorate. So we should assume that, like everyone else assessing the 2016 election, Putin assumed Clinton would win. His tactics were designed to delegitimize her presidency, deepen divisions in the country, and tear at the fabric of institutions that make the United States and the West strong. Specifically, he wanted Trump voters to distrust democracy as an institution by amplifying claims that the electoral process was compromised.

To be sure, the Putin propaganda apparatus has supported left-wing causes as well, for instance, Occupy Wall Street during Barack Obama's presidency.[9] Putin doesn't seem to prefer a left or right ideology. He wants to foment distrust in the ideas this country was founded on—like democracy and capitalism. By promoting fake news about Occupy Wall Street he is trying to stoke distrust of capitalism. He is stoking distrust in our democratic institutions by fueling Trump's claims in the lead-up to Election Day that the vote was rigged. As stated previously, Putin doesn't care on which side of the ideological spectrum their propaganda machine is operating. He solely wants to deepen divisions related to our core beliefs and tear us apart from within. The one exception to his ideological ambivalence is his support for nationalism. He takes any chance he gets to weaken multilateral institutions, particularly the EU and NATO.

The real masterstroke by Putin was to probe, scan, and hack election infrastructure to give Trump's internet trolls and other conspiracy theorists the fodder they would need to spin up one conspiracy theory after another that the fix was in for Clinton's "inevitable" election. We can easily imagine a world where Clinton would have won the election. Instead of the left saying Putin stole it for Trump, rather, the right works itself into a frenzy that this was an elaborate scheme for Putin to steal it for Clinton. Imagine the Russia–Clinton Global Initiative nuke deal[10] coming to light with Clinton as president, and far right trolls are screaming that it is further evidence everything was rigged by Putin in her favor.

POLITICS OF CYBER

What's more disconcerting is that through all the probing, scanning, and hacking of U.S. election infrastructure, Putin learned an enormous amount about how brittle the U.S. election system is. Frankly, he likely now knows more than our own intelligence agencies do about the various election infrastructures throughout the country. Spy agencies like the CIA and NSA do not normally inspect this infrastructure even if asked to do so by local governments because there are so many laws hindering them from doing so. Moreover, When DHS is given access to perform cybersecurity assessments, the scope of those assessments is usually limited and rarely includes all the county and municipal election administration offices where the elections are administered. Obviously Russian intelligence services do not ask whether they can hack local election infrastructure. They also don't care about the difference between state and county systems. Russia hacks each as they see fit. Jurisdictional rules about the separation between federal, state, county, and municipal jurisdictions are irrelevant to a Russian hacker. So while DHS, CIA, and NSA are limited in their ability to learn about the nitty-gritty of our election infrastructure, Russia can probe and research every nuanced detail of our election systems.

Even more worrisome is the likelihood that Russia has penetrated far more state and local election systems than we are aware of. And we have not been able to detect it because many states and almost no local government have the DHS-funded scanning technology deployed on their systems that could have detected it in the first place. DHS undersecretary Christopher Krebs told the House Committee on Homeland Security that it's likely the

Russians scanned every state and that twenty-one is just the number of intrusions they were able to detect.[11] One reason we know about the twenty-one states that were hacked by the Russians is most had the DHS scanning technology deployed on their networks.[12] Nonetheless, the total number of states and territories that had the DHS scanning technology during the 2018 election was forty-six;[13] only forty states and territories had scanners installed during the 2016 election.[14] Even the states that have sensors deployed aren't fully protected. In those forty-six states and territories there are less than one hundred sensors deployed in eighty-four counties and local offices,[15] a number that pales in comparison to the more than eight thousand local election divisions in which voting actually takes place.[16] Of course, it is not politically acceptable to report it that way. Saying there were twenty-one states and leaving out local jurisdictions all together leads the casual reader to believe that less than half of the states were breached and none of the local jurisdictions.

My argument is not that we should assume all of our elections systems were definitely breached. This would be the opposite (but equally incorrect) mistake that some vocal elections officials have made in arguing that, because they have no evidence of an attack, no attack occurred—even though they have no way of detecting an attack on their system. Rather, we should state clearly what we know. DHS was able to detect breaches by Russian hackers in twenty-one states. For the rest of the country, we don't know if they were breached or have any credible way to go back and check. And we will likely never know if they were breached and had an impact on the 2016 election. Further, we have little to no information as to whether the machines that count the votes were hacked and votes were changed, deleted, or added. We also have no way to go back and definitively ascertain if any machines were breached and votes altered.

POINTS OF ATTACK

To understand the vulnerabilities Putin could exploit to undermine trust in our elections, one must first understand the overall election infrastructure system and the key components that comprise it. There are essentially five components of election infrastructure: the databases that contain the list of registered voters, the internal network of an election administrator's office, the voting machines or tabulators, and the public facing website that reports vote totals and the elections software that ties it all together.[17] Each of these has a varying degree of importance to the process, and each has its own unique way in which being hacked would scuttle our democracy. Let's take each in turn.

Voter Registration Databases

The Russian hackers seemed particularly interested in our voter registration databases (VRD). Each state maintains its own statewide VRD system. Most states' VRD systems are hosted on a single platform, with information provided by local jurisdictions (a so-called top-down system). Other states have a bottom-up system, in which the state maintains the "master" database, while the local jurisdiction maintains their own databases to validate which voters can be issued a ballot on Election Day. Any changes to voter registration data are made at the local level, and the local database is periodically synced with the statewide voter registration list. The frequency of the sync varies from near-real-time to once per day. So in these states the VRD system has a myriad of entry points.[18] Take the state of Illinois, for example. The state "master" database is potentially vulnerable to attack, as are each of the jurisdictions' databases. That's a lot of points of entry, and that's just Illinois.

Many in the intelligence community believe Russia may be interested in these databases for another reason: Stealing data from the voter file serves the dual purpose of undermining our elections and building a database of U.S. voters for Russian intelligence purposes. Stealing these databases would enable Russia to research U.S. citizens for efforts other than delegitimizing our elections. While this may be true, I am less concerned than many in the voting rights and privacy community about how catastrophic it is that Russia has this data. In most states this data is easily acquired for free or cheaply with few questions asked. A myriad of political parties, interest groups, and academics are given this data regularly. So Russia can acquire this information with little effort without hacking anything.

However, what Russia likely intended to do by attacking the databases is twofold. First, while others disagree with me, as described in previous chapters, I believe Russia intended to get caught so Trump supporters could cry foul because the election's core data was erroneously considered corrupted and therefore so was the election. This would obviously feed into Trump's narrative that millions of people who aren't supposed to be voting, namely undocumented immigrants, voted. Second, the Russians likely figured out that one of the most disruptive ways to affect the election would be to delete voters from the voter rolls. This reverse ballot stuffing, referenced earlier, is far more disruptive than switching or deleting votes off a voting machine or tabulator. The voter will be disenfranchised in a very public, frustrating, and likely embarrassing way. Namely they show up at a polling place to vote and are told they are not registered there and can't vote. Moreover, if this were to happen dozens of times per precinct, the poll workers would surely notice an unusually high number of people being turned away from the polls. They would become witnesses who could corroborate something was "rotten in Denmark" on Election Day.

This hack benefits Putin because it causes maximum chaos for minimal effort. It also leads to one conspiracy theory after another from every voter who is disenfranchised. News of the Russian hack then fuels the conspiracies, as Putin intended in the first place. Most hackers say penetrating these databases is child's play because, on top of geriatric software, they lack basic "cyber hygiene."

Cyber hygiene most commonly refers to implementing the five most basic cyber controls out of "20 Critical Security Controls" (20 CSC). The 20 CSC were originally determined by top cyber experts at NSA in 2008. The 20 CSC are considered the cybersecurity industry standard, and the top five controls have been proven to prevent 85 percent of known cyberattacks.[19] Unfortunately, almost no organization except the most cyber mature implement all of the top five controls. Big banks, major defense contractors, and national security agencies struggle to implement even the first one, asset inventory, much less all five. So, we can be certain that the top five most basic cybersecurity controls were not implemented on the networks in which the voter registration databases resided in 2016.

In fairness to election administrators, most industries don't know what cyber hygiene is until *after* their industry suffers a major hack. So, voters should pause before picking up their pitchforks and torches. As stated previously, *no one* considered protecting American democracy from an existential threat to the United States as a task in any election administrator's job jar. It's unfair to play "gotcha" about how unprepared they were to defend against a Russian attack. What we should be indignant about are the ongoing comments from election administrators, saying they have "no evidence" that hackers manipulated the election in their jurisdiction.[20] Without the top five controls in place, they don't have the basic technology or practices in place to have any idea what the Russians did or did not do to their databases. Therefore, claims that nothing was changed or altered are baseless.

Many election officials and vendors still claim the voter database is "air gapped and not connected to the internet;"[21] however, the concept of air gapping has been debunked time and again by hackers the world over. This misleading term more accurately indicates the system in question doesn't touch the internet directly. In reality, these databases must touch something that touches the internet or it would be impossible to get new voters into the database or from the database to local election jurisdictions. Moreover, the Mueller Report specifically states the hackers got to these databases by attacking the state election websites showing that even "air-gapped" databases are vulnerable via attacks launched from the internet.[22] Also, as will be discussed in the next chapter, the human race has been connecting our stuff to the internet for the last decade or so in a process called the Fourth Industrial Revolution. Consequently, things that previously were "air gapped" no longer are, because either the device itself or components that interact with the device are connected persistently or periodically to the internet.

The Fourth Industrial Revolution's march is relentless. Think of all the things in your life that are always or sometimes connected to the internet. It has decimated some industries—think taxis—while creating others. Think ride sharing, by connecting drivers to the internet via Uber and Lyft. Voting infrastructure is just another industry to be disrupted by the Fourth Industrial Revolution. The databases and machines were largely cordoned off from the internet 30 years ago. But now "jumping" the "air gap" between the voting machine or database is a trivial affair for an experienced hacker with nation-state resources.

As you will remember from Chapter 1, the best example that debunks the idea of "air gapping" is the above mentioned Stuxnet. Stuxnet was a cyberweapon deployed to cripple the Iranian nuclear program. Specifically, it attacked centrifuges processing nuclear material. These centrifuges were legitimately air gapped. They were encased in concrete vaults. Separated by several feet of cement from anything that touched the internet, they were then *buried underground in the middle of the desert* in rural Iran. Despite all that, hackers were *still* able to deploy their weapon on the network that managed the centrifuges. In an interview with the NSA hackers involved in the attack on the Iranian nuclear program, published in the fascinating film *Zero Day*, the NSA hackers insisted, "We always found a way to get across the air gap. . . . We laughed when people thought they were protected by an air gap."[23]

Clerk's Network and Election Software

The clerk's network is important solely because it gives access to other key parts of the voting infrastructure, for example, the voting machines and databases. The network is just that. It is the system that connects the computers, printers, routers, and other devices in the clerk's office. Of note regarding the election administrator's network is the unique election software tying the disparate parts of the network together. The software is what a jurisdiction uses to ensure voters from the voter registration database wind up with the correct candidates to vote for when they get their ballot, from president down to dog catcher. It also takes results tabulated from the machines and sends the vote totals for each candidate to the clerk's office and ultimately publishes those election results to the official's website on the night of the election. Public security assessments of this software are illusive. But we do know from the

Mueller Report that the Russians accessed the voter registration databases by initially hacking the websites and snaking their way from the website into the voter registration database. That hacking pathway implies two things. First, if the Russians accessed the databases via the website then they got to them via the internet.[24] So much for an air gap. Second, the most likely route for the Russians to have taken would have been through the election administration software that ties the website to the clerk's network to the machines to the voter registration databases. It's possible this software is the weakest link in the chain and no one has assessed its security features publicly, ever.

Voting Machines and Tabulation Devices

We will go into much more detail about voting machines later in the section about the hacker conference, DEF CON. For our purposes in this section, I want to assess the key vulnerabilities. The most obvious threat is that the machine would not accurately record the vote, either by recording a vote for a candidate other than the voter intended or not recording the voter's vote at all. The second, and less obvious, threat, but one I contend may be even more disruptive and easier to pull off, is to break several machines in a particular jurisdiction.

The easiest way for Putin to wreak havoc on an election via a voting machine hack is to insert malware on chips that are already part of makes and models of voting machines. He could then simply break thousands of machines from a particular make or model throughout the country on Election Day remotely. In this way the Russians would not have to account for any unique aspects of a particular jurisdiction. Furthermore, similar to deleting people from the voter registration databases, the benefit of this type of attack for Putin is that it would be a highly visible way to disenfranchise people. Creating several-hour-long lines to vote because of broken machines will also spin up conspiracy theories and delegitimize the winner. Putin's intent seems to be to get caught trying to influence the election. So we lose faith in our democracy. Breaking many machines on Election Day and causing long lines of angry voters would serve his purposes more than a hack to flip votes from one candidate to another that goes undetected.

As we will see with the DEF CON report, parts for voting machines, like most electronics, are made throughout the world. We know the Russians have a long history of breaching the electronics supply chain by infecting parts with malware. So the machine is hacked even before it is taken out of the box by the end user. For years, U.S. Congress effectively banned all hardware manufactured by Chinese telecom companies ZTE and Huawei for exactly this reason.[25] While most of the reported activity about Chinese hacking relates solely to industrial espionage, Russia has also used products manufactured in other countries to attack the supply chain, as they did with the M.E. Doc backdoor, which cost Maersk and FedEx $300 million each.[26] Make no mistake that the line between Russian intelligence hackers and Russian criminal hackers is a gray one at best. It is well documented that the two consistently overlap.[27] More importantly, this is yet another example of this vulnerability not being part of some far-fetched dystopian fantasy. Global supply chain cyberattacks, like all the other hacks mentioned in this book, are happening today.

Public Facing Website

Probably the weakest link in the entire process is the website the clerk and secretary of state use to report election results to the public. As mentioned above, the Mueller Report states the Russians attacked the election websites as an entry point into the voter registration databases.[28] However, it is a mystery to me why the Russians did not hack these sites to report incorrect vote tallies on Election Day. This would have been the ultimate "fake news." It also would have been the easiest way for Russia to tie its fake news and "active measures" propaganda campaign to its attacks on the election infrastructure. In fact this is exactly what Russia did in Ukraine in 2014.[29]

While an election results reporting website was under attack, we would have to assume that people from both extreme factions of the Democrat and Republican parties would spin up conspiracy theories about who was hacking the results and what the accurate result was. Again, this was borne out in Michigan in 2016, when we saw a myriad of conspiracy theories spin out of the rumors that machines there were hacked.[30] It was not until an audit was done of the vote tallies that most of these conspiracy theories subsided. Also, since only a couple states have adequate provisions for automatic paper vote tally audits, there is not a straightforward legal process by which to go about demonstrating an accurate vote count to the public, short of both parties arming themselves to the teeth with lawyers and demanding every ballot be recounted. We saw how that played out in 2000. Furthermore, particularly in states like Pennsylvania, New Jersey, and Georgia, there is no paper record of the vote to recount in the first place[31]; therefore, there is no way to transparently demonstrate to the public that despite the website being hacked, the election administrators eventually arrived at an election result that was accurate.

As with most of the attacks we have talked about so far, this is also not some dystopian fantasy produced by my overactive imagination, but something the Russians have already attempted here and elsewhere. Attacks referenced in the Mueller Report were carried out against election official websites to access voter registration databases. However, through some great reporting by Vox, we have a huge body of evidence showing that in 2018, Russian hackers attacked the election result reporting website for Knox County, Tennessee, elections through a DDoS (disturbed denial of service) attack that shut down the website when results were supposed to be reported.[32] Additionally, in Ukraine, Russian hackers attacked the central election commission and attempted to alter the vote tallies that were reported to the public.[33] The Ukrainian authorities quickly stopped the Russians from hacking the site and reconfirmed the election results; however, as described earlier, by that time the damage was done. Russian media erroneously reported their preferred candidate had won, which spun up conspiracy theories from the losing camp, delegitimized the winners, and undermined confidence in the process.[34]

Do we think Putin really cares who wins or loses a local election in Knox, Tennessee? He certainly does not. That attack was a test to see how we would react to an attack on our election results reporting websites. We should expect an attack like this to occur on the websites of key battleground states in the 2020 elections and begin planning accordingly.

· 6 ·

Here Come the Hackers

Before I delve too deeply into the hackers involved in looking at voting machines and the mindset and story of those who do, it is important to explain DEF CON. DEF CON is a conference started by world renowned hacker Jeff Moss, aka the Dark Tangent. "DEF CON has become one of the world's largest, longest-running, and best-known hacker conferences."[1] It draws almost 30,000 hackers from throughout the world who come to Las Vegas, Nevada, every year to do what everyone else does at conferences: See friends, party, complain about their respective bosses and how screwed up their industry is, get their company to pay for a long weekend in a fun city, and so on.

But there are a couple notable distinctions between this conference and most others. First, many of the hackers are locked in an epic battle against one another in competitions where they both attack and defend networks in challenges that can land the winner not only bragging rights, but also job offers in the half-million-dollar range from blue chip banks, tech companies, or defense contractors.

Also, these hackers are hacking everything one can imagine, from physical locks to smart-city simulations to medical devices to Artificial Intelligence to adult toys. Voting equipment is just the most recent addition. The term "Village" is essentially the room where attendees are hacking a certain type of device and listening to speakers talk about innovative ways to hack that device. While there is a "Voting Village" set up to hack election equipment, there is also an "AI Village" room, where they hack AI, and a "Biohacking Village" room, where they hacked medical devices and so forth in 2017.[2] There are more than twenty of these villages at DEF CON, as well as a million other random activities.

DEF CON is kind of like a macroconference with a bunch of mini-conferences within it. Each mini-conference is represented by a village. If you have ever been to Lollapalooza, Coachella, SXSW, or one of the other mega concerts, it's kind of like that. There are usually multiple stages at the mega-concerts, and some have mostly rock acts, while others have techno and others hip-hop, but it's mostly all tied to music somehow. DEF CON is the same, except everything is related to hacking instead of music. Instead of playing rap, rock, or techno, these guys and gals are hacking vibrators, voting machines, or Volvos. Also, while there are a lot of professionals at the conference, there is just as much purple hair, piercings, tattoos, and the occasional naked person as at Lollapalooza. Even the hackers at DHS who come to the conference to compete against the other hackers dye their hair green or shave their heads or whatever, just for DEF CON. It's wild.

Also like a mega concert, there are a gazillion side events going on. For example,

> some people play capture the flag (a hacker competition where you try and break into an opponent's network and steal their data or some other item deemed to be a "flag") 24 x 7, while many people never touch a computer at DEF CON. Some people see every speech they can, while others miss all speeches. Other activities include contests, movie marathons, scavenger hunts, sleep deprivation, (physical) lock picking, warez trading, drunken parties, a "spot the fed" contest, and official music events. Because DEF CON is what the attendees make of it, there are more events than even the organizers are aware of.[3]

However, one big difference between DEF CON and a typical mega-conference is that the Dark Tangent does not allow *any* corporate advertising, sales, or branding. There are no Microsoft, Google, Raytheon, or IBM banners anywhere. He has cleverly relegated the corporate folks to a separate conference he does three days before DEF CON called Black Hat.

Also, another huge difference between a DEF CON attendee versus another typical Vegas conference attendee is the DEF CON folks do *not* gamble. As world renowned cryptologist and cofounder of the Voting Village Matt Blaze said to me wryly, "It's ironic that DEF CON is held in a city built for people who don't understand math."

Jeff pointed me to the DEF CON website for more background on its origin story. DEF CON.org says the following about the conference:

> Originally started in 1993, it was a meant [sic] to be a party for member(s) of "Platinum Net," a Fido protocol–based hacking network out of Canada. As the main U.S. hub I was helping the Platinum Net organizer (I forget his name) plan a closing party for all the member BBS systems and their users. He was going to shut down the network when his dad took a new job and had to move away. We were talking about where we might hold it, when all of a sudden he

> left early and disappeared. I was just planning a party for a network that was shut down, except for my U.S. nodes. I decided what the hell, I'll invite the members of all the other networks my BBS (a Dark Tangent System) system was a part of, including Cyber Crime International (CCI), Hit Net, Tired of Protection (ToP), and like eight others I can't remember. Why not invite everyone on #hack? Good idea!

The "rules" section of the DEF CON website gives a sense of the ethos of the conference as a mashup of cutting-edge technologists and punk rockers:

> Please refrain from doing anything that might jeopardize the conference or attendees such as lighting your hair on fire or throwing lit road flares in elevators. DEF CON Goons are there to answer your questions and keep everything moving. Hotel security is there to watch over their property. Each has a different mission, and it is wise to not anger the hotel people. Please be aware that if you engage in illegal activities there is a large contingency of feds that attend DEF CON. Talking about how you are going to bomb the RNC convention in front of an FBI agent is a Career Limiting Move!

As one can imagine, a three-day party as futuristic, high-tech, and crazy as DEF CON has its own corner of pop culture dedicated to it. Several of the DEF CON elite are consulting on the recent hit TV show *Mr. Robot*, and thousands of articles, references in counterculture media, and so on are dedicated to it. Notably, there was a huge spike in activity there after the late 1990s movie *Sneakers* came out.[4] For a few years after its release, groupies started showing up looking for hackers to hook up with. That died down after a few years, presumably because as a top hacker from NSA said to me, unlike a rock star, the hackers "wouldn't know what to do with them [the groupies]." I guess "nerd camp" is an apt nickname.

While I would like to think I am hip enough to naturally wind up in a crowd like this, I am not. So I should explain how I ended up working with these heavily tatted-up guys and gals. Mixing with a guy like me is not standard for them. No one I worked with in the political world even knew a hacker before Obama took office in 2009. In the national security world, the relationship between hackers and the security establishment was fraught, to put it nicely. For example, attendees at DEF CON started playing a "game" at the conference in its early years called "Spot the Fed." This was less game than sarcastic civil disobedience. Attendees got sick of the FBI and other law enforcement officials trying to infiltrate their "summer camp" and root out ne'er-do-wells. They made a game out of identifying and outing federal agents.[5] While I was not at all a law enforcement type, as White House liaison to DHS, I certainly seemed to them more like a federal agent than a hacker.

And it's one thing for me to show up at DEF CON and hang around for a day or so. It's another thing to run a "village" there.

Ironically, I have the former head of the largest law enforcement agency in the U.S. government to thank for my acceptance into DEF CON culture. Janet Napolitano was secretary of DHS when I was White House liaison. Part of my job included working with the various boards and commissions for DHS. Upon reviewing Bush administration holdover members of the Homeland Security Advisory Council (HSAC), the most senior and prestigious advisory board for DHS, Secretary Napolitano grumbled something like, "Is there no one besides old corporate white guys qualified for this?" I took that as my marching orders to find women, minorities, and some younger people to appoint. Finding qualified women and minorities was not much of a challenge. Once one starts looking deliberately for them, qualified people emerge pretty quickly; however, finding people who weren't "old" is a bigger challenge.

By the nature of the board, you have to have lived long enough to produce a body of accomplishments of the caliber of the other members. The bar is pretty high, too. The chairman of the board at the time was Judge William Webster. He is the only human in history to head both the CIA and FBI.[6] Finding someone younger than 50 who can hold a candle to that guy is incredibly rare. Nevertheless, because of some good research by the staff at HSAC, I got a mountain of leads for new members. I also had my eye out for someone who had a background in cybersecurity, as I thought it would be good for the department, and for my own selfish reasons since I wanted to meet more people well versed on the topic.

In the stack was a briefing on the founder of DEF CON, Jeff Moss. I remember thinking, "Boom. This is my 'young' guy." I took a stack of briefings for HSAC replacements into Secretary Napolitano. She rejected some and accepted others. When she got to Jeff's brief, she paused and curtly said something along the lines of, "Who the hell is this?" In fairness to her, the picture the research team pulled off the internet for Jeff's dossier made him look like Dieter off of *Saturday Night Live*. On second glance, it looked like his bespectacled face was floating in a sea of black: He wore a black turtleneck and was standing in front of a black background, with just the word—"DEF CON"—hanging above his head.

I made my case why he was a good fit, and it made sense to have one of our "under 50" members be a high-tech cyber guy since he may be privy to more cutting-edge technology. Napolitano was sold pretty quickly, and we sent Jeff and the rest of the prospective members off to be vetted for conflicts of interest, criminal history, or any derogatory information from their background that could disqualify them from the board. Of course, the vetting people came back with red flags on Jeff not because they found any conflicts or criminal history or any other specific problems with him, but since he was a self-described hacker they thought it could generate bad press or attention from Capitol Hill. As political appointees, we *hate* it when the attorneys use their legal review process to make political judgments for us. To her credit, Secretary Napolitano stuck to her guns and said that if the legal review didn't turn up a specific conflict or legal issue, we were free to do what we wanted. The point was also made that we were more at risk of being hit by the press or the Hill for *not* soliciting input from the hacker community. Soon thereafter, I called Jeff to offer him a spot on the task force, and we became fast friends. He suggested I come to DEF CON early on, but I didn't end up attending my first conference until 2013.

Fortunately, however, I was initiated into DEF CON by Jeff, who is like a celebrity at the conference. It would be like walking around your first Lollapalooza with founder and former Jane's Addiction lead singer, Perry Farrell. I remember the first after-hours party we went to, someone had hacked an electronic keg. We also met a guy who was walking around with a backpack on and a scanner device that was sucking up all the data flowing between devices at the entire conference (and likely everywhere else in the casino) so he could "analyze" it after the conference, literally every text, e-mail, and post anyone sent during the entire conference.[7] There are also foreign intelligence services there. The Chinese are the most prevalent. While the full nature of their presence is unconfirmed, multiple unexplained Chinese antennae are reportedly installed throughout Vegas during DEF CON. Presumably to spy on the attendees. There have been discussions about changing the "Spot the Fed" contest into a "Spot the Chinese Intelligence Official" contest, since they have become so brazen in their efforts to surveil attendees it borders on the ridiculous. Most recently, their schtick is to have a male–female couple pose as tourists taking pictures of one another. In fact, they are taking pictures of the hacker in the background of their photo. These pictures would enable the Chinese to build a database of hackers from throughout the world for FBI, Department of Defense, DHS, State Department, and so on. Once they have pictures, they can run them through a facial recognition search of the federal employee photos they have pulled off Facebook or LinkedIn, or stolen from other hacks. Through this cross-referencing, the Chinese can then know which ones work in cyber in departments like the FBI, Department of Defense, DHS, State Department. Then the Chinese can target them for further espionage. It is actually a very clever idea.

There are also the "Wall of Sheep" guys who are trying to hack every conference attendees' phone, tablet, laptop, and so on during the conference. If they successfully hack someone, they put their handle up on a gigantic "Wall of Sheep." This is designed to call out people who go to the largest hacker conference in the world and don't turn on their security configurations. Interspliced in this are DJs playing techno music in different villages, crazy afterparties, and "Goons" in color-coordinated shirts keeping some semblance of order in the chaos.

Goons are hackers who have been around forever and/or done something notable technologically. They are kind of the wise men and women of DEF CON—people who have seen it all before and are so bought into the success of the conference they are willing to work at DEF CON for free[8] for a weekend of their own volition, just for the pride of being a goon. The sages of the hacker world deserve some more focused attention. Later in this chapter I will dive into the background of one of the three hackers most important to the voting village, Harri Hursti. He is one of the cofounders of the village and, like Jeff, pretty famous among the hacker community. He even has a few hacks named after him, the eponymous "Hursti Hacks."

The following is an excerpt of a long conversation I had with possibly the foremost expert on election security in the world, Harri Hursti, and his business partner, Maggie MacAlpine, who is also a reputable hacker and election security expert. It has been heavily edited for clarity and grammar. This is an abbreviated story of how Harri came into the world of hacking. While his story is unique, the general themes are those of many hackers. I had almost no ability to fact check his story except things that are in the public domain, so this should be taken as Harri's word.

Jake: We got all these hackers who stood in line to get into the DEF CON Voting Village this year. I mean, the line was like 100 yards long, and like three people wide, right? Just to get in the village. There was an enormous amount of interest. These were not people who maliciously wanted to flip votes. These are people interested in protecting democracy.

Harri: Or finding truth. They have read about things and want to see if it's the truth.

Jake: Right. And so, it seems to me there is, I don't want to say "political agenda," rather a political level of interest in the hacker community now that seems very apparent to me because of Voting Village, because of other things that I see them being involved in. Not to mention, Russians doing it and everything else. So, a) do you think there really is a real interest in politics from the hacker community, or b) do you think it's just a small group of them that makes it seem like everybody is but really the hacker community isn't interested in politics?

Harri: The hacker community has always been politically active in the United States.

Maggie: I agree.

Harri: It has always been politically active, and it actually is not anarchistic. It's more . . . the hacker community has always had a truth-speaking capability. Or desire. That's why the hacker community has been the first one who does massive forums,[9] get documents out, inform the populace. Politically active is different than picking sides between two parties.

Jake: Granted. And look, I don't want to make what Jane Lute calls the "Christopher Columbus mistake," which is to confuse what's new with what's new to you; however, I will say that what has been done the last two years at DEF CON, linking politics and hacking, it's new.

Harri: We have truly changed the world. Not only the United States but the whole world. When I look back starting on election security in 2005, and now DEF CON the last two years, all actions led us right to the same path. The work is not done yet. But we have made a dent, which no one else has been able to make. I mean, in 20 years, the election activists have failed to make that difference. They contributed, but they failed to make the difference. What the hackers are doing at DEF CON, that is making the difference.

Jake: You think so? Well, let's step back. How did you get into hacking in the first place?

Harri: So, for me, it was really the modems and remote access, which got me started. Before that I was building stuff, and I would start looking into breaking stuff, but when modems happened, then it all changed for me because now it's a question of, "Oh, I can remotely control this." Also, I don't need to pay for long-distance calls, ha ha ha. So, '81. That's the year I got started [hacking]. Late '81.

Jake: Really?

Harri: I built my first computer in '77. And '81 was the time when I went to the remote access world.

Jake: Ah, you're older than I think you are.

Harri: No, I started younger than you think I did.

Jake: Ah, that's funny.

Maggie: So Harri was nine when he first started hacking and fourteen when he had his first company.

Harri: Yeah, so I wasn't interested in computers at all. I was interested in stars and astronauts. and then there was a couple of guys who picked up on my skill in math. I knew I would never be able to do math;

however, physics is math for people who cannot do math. Ha, ha, ha. You know, infinity minus infinity is a real equation, which in real math is not possible. So, there were a couple of people who spotted me and then started giving me access to computers. Then modems and remote access and telephone networks became the transition point for me.

Jake: And it was because of remote access?

Harri: It was because of remote access. That was the point when everything changed for me.

Jake: So what do you think that is? That's really interesting. I mean, I have ideas on why that is. But what do you think it is? Your self-diagnosis. Ha, ha, ha.

Harri: It's just the whole idea of the fact that you can use something, which is remote. I've always been interested in radio. I was always interested in telephone networks. I did word processing. I wrote programs. But that was not my thing. I mean, I did it, I was good at it. Even when I was young. But it was not something I fell in love with. Then modems and remote access came around. That changed everything.

Jake: When did you find out that there was anybody else who could do this or gave you shit about it or . . .

Harri: Immediately.

Jake: Oh, really?

Harri: So '80, '81, the guy who ran IT for the largest bank of Finland, he built a bulletin board system. That was when it all got started for me because computers were very expensive, and I was at the time working for a company which had customers returning computers they had bought. I had access to components which probably would cost an arm and a leg. This is how we really got everything started.

Jake: And once you got a modem, you could travel around the world.

Harri: True.

Jake: So then you start finding other people around the world who are into the same shit?

Harri: No, international calls were really not working very well, but I found people from Finland. Actually, 1983, we were starting a new UUCP, Unix-to-Unix Copy Program, e-mail system called EUnet, which later I cofounded when the technology, originally used for hobbyist access,[10] became a business. So, in 1993 we bought the association and altered and turned it into a business. Ten years later.

Jake: Oh, yeah?

Harri: Yeah. That was when the new UUCP turned into the internet.

Yeah, it always has been for me really about communication. Also, if you think about the memory card, in a way, a memory card is also communication. It's a device inside of another device. So the underlying idea is communication. So, the whole thing then leads to the situation where a publishing company wanted to start a computer magazine. They contacted me. I sold the idea to the publishing company that if you subscribe you should get a free email service.

Jake: Oh!

Harri: There was no e-mail service. So, they didn't believe I could do it. I made a specification to hire the biggest system integrator in the country to develop it. It turned out to be very complicated. They developed the thing, but it turned into a complete disaster. So, two years later, they made a decision that I could buy a mainframe. I put in my own team and developed a better version of that program.

Jake: And so how old were you at this point?

Harri: Sixteen? Fifteen? Sixteen.

Jake: You were managing adults, of course?

Harri: Yes. Which was . . . very challenging because adults had adult problems, and I was like, "No, don't, we just need to do this project, you know . . ."

Jake: Ha, ha, ha. You didn't start finding people around the world until what, the late '90s? Because, I mean, you really couldn't until then—

Harri: I started to find people around the world in '86, '87. Because at that time there was no real-time internet in Finland at all. The nearest place was Denmark, in Copenhagen. We then were able to do international calls to Denmark and use BITNET from there. I used e-mail since '84, which, by the way, might have taken a day or two round-trip, because they were relayed over twenty sites.[11]

Jake: Okay, it's 1984, and you're talking to people in other countries, and is it all kind of about the . . . the same stuff: modems and—

Harri: There's one more thing which happened. I was transferred from the school I was initially in by my parents. When I got to the new school, I immediately got beaten up and I was in emergency care. My knees were broken. I was very, very small. The smallest of the class. Smaller than any girl. I got beaten up by the son of a prominent politician. I had horrible years because my parents refused to transfer me out of school because this is the school where all top politicians' kids are. Eventually what happened was that, because of my computer skills, I got the attention of the secretary of education. She learned about my situation, and she made a deal with me. I will become one of the representatives of the government of Finland in

UNESCO, in the United Nations. In return I choose any school I want and I will get transferred there. I was transferred to a university. She had direct control to tell the headmaster that my not being present in class was not a concern. But that was yet another time when I was given more access to the biggest computers in Finland. One thing which I hate so much is one of the guys who beat me up is now an adjunct professor here in New York. He told me that I should be grateful for what they did to me because I wouldn't have been transferred to this university and get all these privileges without them beating me up and getting me sent to the emergency room. Literally my knees were broken. I used to use canes for over a year before I was able to walk without canes. The surgery really didn't work properly. It was an experimental surgery. For over a year I couldn't walk without . . . what do you call those? Those things which are . . .

Jake: Arm braces or something?

Harri: Yeah. the sticks which are, the whole arm goes in them you . . . anyway, horrible time.

Jake: After high school?

Harri: I started a company where I developed software for the military. In Finland military service is mandatory. Since I had the school-violence broken knees, I was in a position which everybody would have wanted to be. I was a person who doesn't need to serve; however, I was convinced that for business, it would be good if I volunteer. Mandatory service starts when you are eighteen. When I was seventeen, I elected to volunteer. I was assigned to the general headquarters to the department of data processing and, ultimately, data processing and signal intelligence. We broke so many laws even getting me into the military. We broke the Paris Peace Treaty with my assignments, ha, ha, ha, a number of times. Being in the military led me to encryption. I really wasn't that interested in encryption, but I was given a project to build communications the Soviets cannot penetrate. But I really didn't know much about encryption. That project is what pushed me to open source. I very quickly realized that if this is done the "military way," it was all very secret. That would put my personal safety at risk (if I was one of few people to know all the codes). I insisted that the software has to be made public and that the algorithm has to be made public. The only secret would be the key, which changes daily or hourly, not how the software works. I managed to get most of my military programs to become open source because we have to eliminate ourselves as people who know something other people don't know because we want to die in old age.

Jake: When did you leave the military?

Harri: I came in '87, and I left in '89.

Jake: You were nineteen. So after you get out what happens?

Harri: I left and started the first Pan-European internet service provider. That was really an easy choice because now I got back to—

Jake: Modems.

Harri: Modems.

Jake: And remote control.

Harri: Yeah. Well, I did a little bit of that in the military. Let me rewind a little bit. So '87 to '89, the Finnish military was at a crossroads about the fighter jets. They had two choices. One was the MiG, a Soviet-made jet, or the McDonnell Douglas-made (later acquired by the Boeing Company)[12] Hornet. The whole Soviet Union network ran with the DECnet,[13] the American-made DECnet.[14] That's how Soviet Union network worked.

Jake: Really?

Harri: Yeah! Always. We bought the Russian jets equipped on DECnet. So initially we used DECnet. If we bought Hornets, those were on the TCP/IP [Transmission Control Protocol/Internet Protocol] Network. One of my last jobs was the transition for Finnish military to the TCP/IP Network. That was one of my last jobs that launched me to the internet. It's a relatively little-known fact that all of the Soviet Union was ran on an American corporate network. Ha, ha, ha, ha.

Jake: How did the Soviets allow themselves to do that?

Harri: The Soviets were very good in copying certain computers. They learned very early. They copied Apple computers, and then they copied Windows. They never really copied IBMs. It's a question of what they learned to copy. It was a known thing that DEC [Digital Equipment Corporation][15] was feeding information to the Soviet Union. There were a lot of scandals around that. DEC was really massively leaking information.

Jake: When did you first hear the term "hacker"?

Harri: Now, "hacker" has meant many things over the years. "Hacker" was originally a tinkerer. Initially "hacker" is for those who do illegal telephone calls. Like Blue Box, which was made by (cofounder of Apple) Steve Wozniak and other people. Really "hacker" became a bad word only when Kevin Mitnick [an infamous hacker from the '90s] made it bad. Before that "hacker" was a good word. It was a nerd of sorts. It was a tinkerer, experimenter, and stuff like that. There was a bad word, "cracker," which was people who were "cracking" software.

Jake: It's so funny because that's a derogatory term for white people in the United States. "Cracker."

Harri: It really was a term from the '80s to the '90s. "Hacker" became a bad word, I would say in '90, '91. That's when it started to have a bad connotation. I don't remember which year, but the movie *War Games* [came out], that also accelerated the idea of a hacker.

Jake: That's '83.

Harri: That really is the time when "hacker" turned from a good word to bad word.

Jake: Okay, so early on you're just meeting other Finnish hackers, and you're meeting them at work and social situations.

Harri: Back in those days e-mail addresses were called a "*bang* address." They had parties called either "bang parties" or "@ parties" because that was the parties for people who have never seen each other in person. They have their e-mail addresses displayed (like on a name tag). These were parties where you can meet the people who you have sometimes talked to for years only online.

Jake: That's awesome.

Harri: Basically, the problem in those days was to find out where the meetups were and which meetups are worth going to.

Jake: So how did you figure out which ones were worth going to?

Harri: In those days there was a thing in Germany called the Chaos Computer Club (CCC), which is still around. CCC was a good clearinghouse of information. So, that was very important to find out what things Chaos Computer Club is looking at. Chaos Computer Club at that time was a zero-criminal organization.[16] Oh, and the one book which changed the world for hackers was Clifford Stoll's *Cuckoo's Egg*. We actually know the person who saw the "assisted suicide" of the East German hacker who hacked Lawrence Livermore. He committed suicide by pouring gasoline over himself and lighting himself on fire. That was assisted—

Jake: Why would you pick that as your way to die?

Harri: No, no, no it's "assisted suicide." Meaning, that guy . . .

Maggie: He was assassinated.

Harri: He was killed in front of audience.

Maggie: Yes, Harri's sarcasm is sometimes lost. This man was killed to make a point.

Jake: Oh, wow.

Harri: This whole thing used Tymnet,[17] so BITNET[18] was run by IBM. So Tymnet was run by McDonnell Douglas.[19] They were thinking they were hacking the national laboratory, but they were hacking Lawrence Livermore University. So that's why Richard Stoll created documents which were all garbage but looked like secret documents. Every single time they tried to download them it created the artificial static. They never got the downloads working, but he was able to keep them online.

Jake: That's pretty awesome. Yeah, I'm trying to remember. I read this forever ago. Like in high school or something.

Harri: The author, Richard Stoll, in the United States was literally the first hacker catcher . . . he was, he was also an astronomer, like I wanted to become.

Jake: Oh, really.

Harri: Yeah, he was not a computer guy, but nobody else had ever caught an international hacker trying to steal data. He was the first one. So he became a guru overnight. People were calling him asking, "How did you do that?" And he says, "I'm an astronomer. I like stars." You know? "What the hell are you talking about catching hackers?" (Ha, ha,ha.)

Jake: Huh. So you were just kind of building product and communications companies?

Harri: Yeah.

Jake: And then you met Matt Blaze in '07. and he's the one who invited you to DEF CON.

Harri: No, no, no it actually was Mouse [her hacker handle], Sandy Clark.

Jake: Oh, she's amazing.

Harri: I knew Sandy well before I knew Matt Blaze. Maus is a very interesting character because her skills are hardware. But why she is very useful is she is a team-builder. She is a person who can take a team and make that work. Maus as herself, she's good, but she's not exceptionally good. But Maus knows just about everyone.

Jake: Yeah, my first DEF CON, maybe Jeff Moss [founder of DEF CON] asked her to babysit us. I don't know. Maybe she just kind of realized we were out of place. I bet I spent a quarter of my time with Maus.

Harri: Oh, really?

Jake: She just came up to us and was like, "Who are you? What are you doing? Why are you here? And, oh, you're in politics. Do you hack politics?" and I'm like, "Well, I've never thought of it that way, but sure." She's just being nice, you know, trying to make me feel I fit in even though I didn't. I totally get your point about team-building.

She's trying to make me feel like I was part of the team even though I clearly was not. I was a total outsider looking in, you know?

Harri: Yeah, that's who she is. And she has been around forever. She's actually running a program where she is trying to convert criminals to be honest cyber workers. She asks the criminals to provide everything they have done as hackers and rates that as if it would have been part of your degree. Then basically you get college credit for everything you have done, which is an excellent idea.

Jake: Wow. How did you meet?

Harri: From '84, there was no national, commercial internet access. In Europe I went around founding this EU internet company. I think Maus at that time was living in Austria. I probably met her for the first time in Austria.

Jake: Huh.

Harri: Because of my role building all the infrastructure, I was in the United States all the time, at the North American Network Operator Association (NANOA) meetings. At one point, the network I ran was the first-tier network in the world. First-tier network meaning that there is no transit operator [and information is exchanged with other first-tier networks through peering agreements, with no fees].[20] And we negotiated a deal with the IIJ [Internet Initiative Japan] in Japan, and we got the Asia first-band connected. No one in the United States, no one in Europe, had been able to do that. So that was when we were the first-tier internet service provider in the world.

Jake: It was at these network association meetings you guys all started meeting each other?

Harri: I was on the IETF[21] [Internet Engineering Task Force], where all the internet standards were being set. Basically we were building the internet. I was all the time in the weirdest places for meetings. A lot of connections were made in those days.

Jake: You know I've been thinking about—total side note—what's really interesting that's gonna happen in our lifetimes, is we are gonna connect the last 3 billion people on the planet to the Internet, and we're going to connect essentially all of our stuff to the internet. And like, once that happens, it's not gonna happen again. I mean, new people will be born and new stuff will be made, but they will all be immediately connected to the internet.

Harri: Yeah.

Jake: And new stuff will be made, but if there's a reason to connect, it will be—

Harri: And even if there is no reason to connect it, it will be connected, yeah.

Jake: Right. The connectivity will be built into all the stuff and all the people.

Harri: Yeah.

Jake: And once that's done, that's never gonna happen again.

Harri: True.

Jake: You know? I mean, that's gonna happen in our lifetime. It's crazy, it's totally crazy.

Harri: I remember when we were celebrating the first 5 million computers connected to the internet. I remember when we were celebrating first 10 million computers connected to the internet. I mean, that's how young the internet was when I got involved.

Jake: That's why I raised it. You were one of the people there when "they" said, "We are gonna start connecting stuff to the internet. Even all over transcontinental connections from Japan to the United States and Europe, etc. Now, in your lifetime, all our stuff and all the humans on the planet will be connected to the internet. Okay. This is really interesting, so you get into tinkering, right? Because of modems. Computer systems aren't what does it for you. It's the remote control of machines, connectivity.

Harri: Yeah, I was tinkering with electronics, but really where my passion started was the remote access.

Jake: Right, then, you start meeting all these people at these network admin or network connectivity conferences or whatever, where you are connecting Japan and Europe or the United States and Europe to the internet. After you guys get into this idea of connectivity, now you're having parties for connecting the next 5 million, the next 10 million computers to the internet. So that's kinda the progression. Is that right or no?

Harri: By the way, did I tell you that I made the first three connections to the Soviet Union to get on the internet. So the first—

Jake: Wait, the first what!?!

Harri: The first three connections of the Soviet Union to the internet. The first connection was a UCP [Universal Computer Protocol], so a dial-up type of connection. The second connection, which was approved by the U.S. government, was an organization called DEMOS.

Jake: Davos?

Harri: DEMOS. D-E-M-O-S. That was a think tank. It was a known KGB front of 400 researchers. Their connection was 2,000 characters per second from Helsinki to Tallinn, because they had a fiber line to Tallinn. The weirdest thing happened. At the time the internet was controlled in the United States by an organization called IPC. And they gave a connection to us, asking, "Can we get a two megabits second line to Leningrad?" Which two megabits per second was an amazing amount of capacity. I checked, and I found out there was a hidden fiber. Sprint delivered me two two-megabit lines to Helsinki. I delivered the line from Helsinki to Leningrad. I was like, "Where is this line going in Leningrad?" And it went to the Kurchatov Institute of Atomic Energy.[22] That is the only nuclear weapons laboratory in the Soviet Union. I said, "Well, that's very interesting. Ha, ha, ha, ha, ha." It turned out that there was an actual agreement between the U. S. government and the Soviet Union that it was a real-time data feed for certain preferred people. So nuclear material production, nuclear records after Chernobyl, stuff like that was sent to the United States. Now, fast-forward, and this is funny because you can find the actual logs for the coup d'état in the Soviet Union.

Jake: The what!?!

Harri: Coup d'état. Gorbachev is in house arrest in Sochi. There are three guys actually in power. Yeltsin hasn't yet risen to the occasion. Everything goes dark in the Soviet Union, except that original UUCP line stays up and keeps on pulling data every fifteen minutes or half an hour to the United States. Of course, this line is well-known by the KGB. So, the fact that it stayed up was not an accident. You can find a log by "Scofield" [Harri's hacker handle] and "See You Again in Next World Crisis" because when I closed the channel[23] I said, "This crisis seems to be cooling down, see you again in next world crisis," and I started shutting it down.[24] That was about a day after the coup ended. Yeltsin had assumed power. Gorbachev returned to Moscow. It was a really hectic few days because basically we were operating that thing twenty-four hours a day. I mean, the reason the whole thing was there was exactly this type of a crisis. That was really hectic.

Jake: That's incredible. Well, listening to you now for a while, it seems to me like it's less that there's moments and more like there's eras. Part of it is just age. Things that happened in high school are the types of things that happened to people in high school. At the same time, there is Cold War–era technology in the '80s. Then all of a sudden now modems are a thing. Then there's this post–Cold War era in the '90s where we are connecting the whole world to the internet. During that era you are getting connected to this broader community of technologists and would-be hackers all over the world. Then we are in today. We are doing all the shit that we're doing right now with DEF CON. How would you describe this era versus those previous eras?

Harri: Are you asking this era as the current wave of foreign influence? What is the definition of this era?

Jake: Yeah, good question. I don't know. I'm kind of asking you. I'm trying to define it myself. Honestly, we are the ones describing it to the media and policymakers, and I don't even know what it is myself. Before we get into that, there's a decade between the Cold War ending and September 11. How do you characterize that?

Harri: That's true. That's the period of hope.

Jake: Well, I mean you guys were connecting the internet everywhere.

Harri: We were doing that. Things got easier. Less visa requirements. Less hoops everywhere. More global. Bill Clinton actually officially stated that the second-highest priority of the CIA was corporate espionage benefiting American companies.[25] He actually changed the CIA's mission statement! The second-highest priority is corporate espionage that benefits American companies. That tells you that the world was in a good place.

Jake: Right. If that's all we gotta worry about, then we're doing okay.

Harri: Then September 11th happened, and that set things on a new path toward another Cold War.

Jake: With Russia or with whom?

Harri: With everyone.

• 7 •

"You Have to Sit on Those Boxes"

"You have to sit on those boxes."—Franklin Delano Roosevelt, speaking to Lyndon B. Johnson after the latter lost his first Senate run in 1941, because he allowed his opponent to stuff more ballot boxes

In early 2011, I found myself back in Chicago. I had just left the Obama administration, and I continued advising DHS as a consultant. But by 2014, I was ready to go back into the administration. Our kids were not infants anymore, and Jena was fine with moving back to Washington. I had two choices: I could see if some of the people I knew in the Obama administration could find another position for me or try and help Hillary win in 2016, and see if I could get hired in the early years of her administration. Obama administration people were leaving in droves to try and leverage their relationship with the current administration leadership while their personal capital was still high.

After Obama was gone, all bets were off as to whether any of the former Obama people would have access or influence in the next administration. So the people in Presidential Personnel were dying to find people to fill key posts. I thought I had a decent chance of going back in, but had the same concerns everyone else who was leaving did about what my market value would be if I stayed until the end of Obama's term and was forced to leave when the Clinton people came in. Many people at the end of the Obama administration were under a delusion that the Clinton people would keep them around if she won. As I had never been through this before, I asked around to see if entering the Obama administration at the end and then being kept on by the Hillary Rodham Clinton (HRC) administration would be possible.

That hope was quickly dashed when I spoke to my business partner, Chris

Burnham, who was in the Bush administration. He quickly gave me a dose of reality, saying,

> No way. I remember the first day of the H. W. Bush administration. They asked for all the Reagan presidential appointees' resignations on day one and accepted every one of them. Everyone was whining and like, "Oh but we are all on the same team," and the Bush people were like, "Sorry. New president, new team." And by the way that was in the Bush and Reagan administrations. They actually got along, and their respective people generally respected each other. How do you think this will play out between the Hillary and Obama people?

He was obviously alluding to the well-known fact that while it seemed Obama and Hillary had come to respect one another throughout the years, their respective teams hated one another. The fights between the HRC State Department staff and the Obama White House staff were legendary by that point.

I actually prided myself on knowing I had a hand in getting one of the only Obama campaign staffers into a presidential appointment in the State Department in the first two years of the Obama administration. I was deputy director of the national security agencies review for the presidential transition team in 2008 and 2009. In that capacity, I managed the support staff for the team of experts who study the current state of affairs in each agency so as to brief the incoming president, cabinet secretaries, and senior staff.

While most of the teams conducting the review of each national security agency generally met at the transition office and were grateful for the administrative support the twenty or so former Obama campaign staff provided, the State Department team virtually never met at the Transition HQ and not only refused administrative help from our campaign staff, but also flat-out refused to even allow the campaign staff assigned to support them to enter the State Department building in Foggy Bottom. There was one notable exception, however: Jim Steinberg, HRC's soon-to-be number two at the State Department, was above the petty slights being exchanged between the HRC and Obama teams. He occupied an office at the Transition HQ and was pleasant and kind to everyone I saw him interact with, including the young former campaign staff that were providing administrative support.

For the young campaign staff supporting the senior leadership, time on the transition is like an extended job interview. While the staff for the other agencies were ingratiating themselves into the incoming leadership in their agencies, the State Department staff were getting nowhere and thinking they would not have a job once the transition was over. One of those staffers was Laura Updegrove. Day after day, she became increasingly concerned she would be out of a job the day after the transition ended. As her anxiety increased, I saw an opening for her with Steinberg once he decided to office out of the transition.

"Just go stand outside Steinberg's office all day," I said.

> Eventually the guy is going to need a cup of coffee or copies or whatever. Once he gets used to you being able to get stuff done for him, he will keep asking. Then hopefully he will bring you over to State with him, as he knows you are an Obama campaign staffer and will get approved by Presidential Personnel quickly.

She stood outside his door for hours. Periodically, she would come into my office and say, "Jake, I have been standing there for three hours, and he hasn't asked for anything. He is going to think I am crazy."

I told her to continue stalking him, and eventually things worked as planned. Steinberg eventually needed help from Laura, and she did an outstanding job performing even the most mundane tasks. When he went into his role as number two of the State Department, he brought Laura with him. Almost every other campaign staffer who worked for Obama in the primary was blocked by the HRC team for the first two years.[1] In fairness, any HRC primary campaign staffer who had not worked for Obama in the general election had an infinitesimally small chance of breaking into the Obama administration, too.

Laura's experience, coupled with Chris's anecdote about the Reagan and H. W. Bush administrations, made me decide that if HRC won and I wanted to be involved in the national security establishment for the next four to eight years, I would be better off forgoing a position in the last two years of the Obama administration and try to help the Clinton campaign instead.

I had always hated fund-raising, but that was the most highly valued activity you could do for the campaign short of quitting your job and moving to work full time for the campaign somewhere like Iowa or New Hampshire. So I became a fund-raiser, or "bundler," as they call it in federal campaigns, because you can't just write a huge $50,000 or $1 million check to the campaign. Federal Election Commission rules state that to raise $50,000 or $1 million or whatever amount, you must collect, or "bundle," checks from many people of no value greater than $2,700 each per individual donor. Even Warren Buffett and Bill Gates can only give a particular federal candidate's political action committee $2,700. There are myriad ways around this rule because of the Supreme Court's "Citizens' United" ruling that are too complex and numerous to go into here. Bottom line is I had to call, text, and otherwise harass lots of people for checks of $2,700 or less.

My first foray into the fund-raising arena for Clinton was to try and organize some people I knew in the cybersecurity industry to host cybersecurity-themed fund-raisers. To my knowledge, these were the first "cyber fund-raisers," at least on the Democratic side of the aisle.

The second one we did was called the "Hackers for Hillary" fund-raiser, and it happened at Black Hat, the sister event to DEF CON. Black Hat was the more buttoned-up, professional, and corporate of the two conferences. As such, Jeff Moss, who founded Black Hat and DEF CON, and I figured it would be a better place to hold a fund-raiser because we assumed people there would have more money to donate. This fund-raiser was held on August 3, 2016, at a Mexican restaurant near the Black Hat conference.

I didn't realize it at the time, but this fund-raiser was part of a series of events in 2016, including the Russian hack on the 2016 election and Trump's subsequent denial of Russia's role in the attacks, that injected politics into cybersecurity. As Harri Hursti said, while many hackers were political, for decades going back to the 1970s and 1980s, cybersecurity and hacking were spoken of in mostly technical circles with few policymakers discussing or even understanding where "cyber" began and ended. Even fewer people knew what to do about it, save a few military personnel and spies. Later, in the 2000s, nontechnical policymakers, especially from the national security world, began to talk about the national security policy implications of cyber. Policy wonks at think tanks and senior government institutions who didn't know a firewall from a zero-day began engaging in increasingly consequential cyber policy debates that were often devoid of sophisticated technical understanding from previous years.

Furthermore, highly technical cyber experts were engaging in far-reaching policy discussions with little academic training or hands-on experience with such hot-button issues as human rights, war and peace, or diplomacy. While there is a long tradition of policy and laws playing catch-up to technological advances, cyber was different in the sense that so few policymakers understood what the internet actually is or what cyberattacks are capable of. Moreover, this was unfolding as the Fourth Industrial Revolution was surging forward and connecting almost everything humans made to the internet, and connecting the last 3 billion humans to the internet as well. Just to make the collision of rapid technological innovation and policymaking more complex, the events of 2016 added politics into the cyber equation in a very public way.

Moss and I were told that very few people would be interested in the fund-raiser because the cyber world eschews politics, especially "Big P" partisan electoral politics. Many people from this community would generally be considered apolitical or libertarian at best. A fund-raiser for a political candidate was anathema to the zeitgeist of the people in this community; however, while

Moss especially got significant pushback for being associated with the event, there was enormous interest from people who both donated and attended. The number of people in the room quickly exceeded capacity, and we had to use a spillover room to accommodate everyone. We only raised about $30,000; but considering we didn't have HRC, Bill, Obama, or the First Lady, and this was the first fund-raiser of its kind, we were surprised by the outcome.

To have cyber play this public of a role in a political campaign was unprecedented. Cyber was a non-issue in every election before this one. It was only mentioned once in the presidential debates in 2012.[2] No cyber activists groups came out in support of either candidate and neither candidate got much coverage, good or bad about anything they said or did in relation to cyber. Before 2012, cyber as a policy topic wasn't even *discussed* by most presidential candidates' policy advisors, much less as a campaign issue. In this election however, one candidate (HRC) was being attacked incessantly for what was in part a poor cybersecurity decision, putting her government e-mails on a personal server. Then the opposing candidate, Trump, calls on a foreign government to hack presumably U.S. citizens' networks to find said e-mails. Fortunately for us Trump said this only a week before our fund-raiser. So the cyber people at Black Hat were even more tuned in to what was going on. That is a remarkably fast evolution for a topic, in this case cybersecurity, to go from literally no relevance at all in presidential politics to a front and center campaign topic that is being discussed by everyone nonstop. We had been planning the event for about a month before Black Hat, but I assume Trump's call for the Russians to hack HRC's e-mails helped with attendance.

As usual the media was feeding the political flames. We knew the press would be interested in the event. In fact, I believe it was the press who dubbed the event "Hackers for Hillary." But the level of interest was far beyond anything we expected. Because the interest was so high and the Russian hacking issues were starting to gain traction, I called Robby Mook, Hillary's campaign manager. I asked him what talking points he wanted me and the other speakers to use or, more importantly, not use at the event.

Even in these early days of the revelations about Russian influence in the election, Robby was seriously concerned about the potential implications for the HRC campaign. He connected me to Glen Caplin, then a spokesperson for Clinton, who spent a considerable amount of time on the phone with me on my way to the airport headed to Las Vegas for Black Hat. A real communications pro, if not a person steeped in cyber policy, Glen coached me through the key positions of the campaign. Hacking elections? They were against it. Fake news promulgated by Russians? They were against that, too. Highlighting Trump campaign ties to Russia? They were for that.[3]

The last thing I wanted to have happen was to turn this well-meaning fund-raiser into a press conference where I said something stupid to draw the ire of the HRC campaign. Glen was great and got me in a good place mentally to make some key points and not take too many questions to minimize the opportunity for gaffes. I had my staff do some research on what Russia had done to interfere in elections in Eastern Europe and Central Asia. I tried to clearly make the point that we need to assume anything they did over there, they will try here.

The event went off gaffe-free. While we hit our modest fund-raising goal, I was most struck by how many people crammed themselves into the event and even into the overflow room, where they couldn't hear any of the speakers and were seemingly there solely to talk politics over a beer with other conference attendees.

The interest from the media was astounding. At one point the attendees began complaining, loudly, that so many reporters had elbowed their way into the front of the event that none of those who bought tickets could hear the speakers. In fairness to the reporters in attendance, the event was in a loud bar, and they were trying to make sure their microphones could pick up what we were saying.

The revelation that the cyber community was actually very interested in partisan politics, and the tech media was even more interested, stuck with me. It seemed those of us trying to understand this entire ecosystem had been missing something. The thing we had been missing was the growing injection of politics into the cyber world.

After DEF CON, I went to Clinton's Brooklyn campaign headquarters to thank Glen and see Robby and my friend Teddy Goff, who was overseeing digital communications efforts[4] for the campaign. Teddy laughed at the feeding frenzy the tech press had with it, and Robby thanked me for raising the money. I mentioned to Robby that person after person at the event had been giving me dire warnings about how proficient the Russians are at hacking and how vulnerable our election infrastructure is. Moreover, to a person, they were offering to help the campaign in any way they could to either stave off an attack by Russia or help protect local election infrastructure. The latter, of course, Robby had no control over.

At that point I had known Robby for more than a decade, since the 2004 campaign for John Kerry, and I believed I could tell when he was genuinely interested in something you had to say or was just politely agreeing with you. He seemed legitimately interested in this offer of help from the hacker community and said, truthfully, that he didn't know how to use what were essentially cyber volunteers, but he wanted to keep discussing it since it could prove useful. Bottom line is that I am certain Robby was taking the Russian threat to the campaign very seriously.

I also knew Robby was concerned about Russian hacking because of a conversation we had at the Democratic National Convention a few weeks earlier. I was a whip for the convention, and Robby was running the show, along with Charlie Baker, an old political hand who had worked on something like eight presidential campaigns. One morning I was eating breakfast with my family in our hotel, and Robby stopped over to say hello.

Trump had just executed what we later realized was a typical Trump tactic of yanking positive media attention away from his opponent by saying something so outlandish it compels the press to cover Trump instead of his opponent. In this case he had just made the aforementioned comment that, "If Russia finds Hillary's deleted e-mails, they will be rewarded greatly."[5] He later clarified that he actually said at different points in his speech that Russia would be rewarded greatly by "our media," not his campaign or his administration. The fact that one American presidential candidate would ask a foreign power to hack another presidential candidate had the desired effect of working the media up into a lather and taking their focus off the DNC Convention. It seemed only Trump understood that the media discussing whether he had crossed the line of committing treason didn't matter. It only mattered that they were talking about Trump and not Hillary.

I said, "This Russian hacking stuff is crazy isn't it?" as the clip of Trump's speech played in an almost continuous loop on CNN a few feet from our table.

Robby said something along the lines of: Yeah it's nuts. When the convention is over, send me an e-mail and I will hook you up with our folks dealing with this. Maybe you and your guys can be helpful to them.

As stated above, our fund-raiser at Black Hat was a week after the Democratic National Convention, and before I could take Robby up on his offer, I had to get back to D.C. to get some work done with my clients. I had been ignoring my day job for almost a month between DNC Convention and Black Hat so I needed to get some real work done.

While back in D.C., I had two meetings I immediately should have realized foretold disaster. In one meeting, I spoke to a senior official of the analysis division at DHS who was working on an assessment of voting infrastructure risks for DHS and ultimately the White House. He and I met for breakfast at a diner in Arlington, Virginia, where many of the national security elite meet to evade the prying eyes of reporters or partisans who seem to be lurking around every corner in the District.

We were both seeking information from the other about the election and our security leading up to November. I was there to find out if they were considering every aspect of the election the Russians could "hack." I wanted to make sure they realized that accessing the machines and vote tabulators and changing vote tallies was just one of many ways Russia could influence the outcome of the election.

Similarly, he wanted to pick my brain about whether his team was missing anything in its assessment. He mentioned that the political appointees that were left at DHS had little to no campaign experience. Most of the staff from the campaign in 2008, like me, had left the administration years ago. So the political appointees knew little more than his team, the members of which had never worked on a campaign or seen how an election is administered outside of voting themselves. On top of that, the political appointees were being advised to stay as far away from the election security work as possible. One poorly worded e-mail or ill-timed decision could damage the legitimacy of the entire election and send the conspiracy theorists into a furor.

I was kind of shocked that they didn't have anyone with campaign experience involved in their risk assessment process. I thought they would need an entire team of people who worked on voter protection for us in '08 or '12 and several local election administration officials. I explained that a bunch of analysts had no chance of getting up to speed on the nuances of the unique aspects of the election administration processes in thousands of jurisdictions in more than a dozen battleground states, not to mention the nonbattleground states.

He said the administration was terrified of any activity being perceived as an attempt to influence the election for Hillary. He even said he preferred to meet with me in a diner rather than his office at DHS because he didn't want anyone to see us together at DHS. I was a known political operative with strong ties to DHS who was also raising money for Hillary. If anyone saw us together, that could set off a host of conspiracy theories about the HRC campaign trying to impact the national security officials responding to the Russian threat. Obviously, this was ridiculous. Aside from raising money for the HRC campaign, I wasn't involved in the campaign and was not working on election security at DHS.

But it didn't matter what the reality was. Perception was the thing everyone was most concerned about. Hillary was well ahead in the polls, and everyone was "sure" she would win. No one wanted to be responsible for some ill-advised action around election security that would inadvertently influence the outcome of the election. In fact, my colleague was a Republican, so the irony was not lost on him that some may think he wanted to help HRC win when he planned to vote for Republicans. He was in a tenuous situation at best. Understandably, he was being incredibly cautious.

Regardless, I was still at a loss as to how this senior official's team at DHS could accomplish an effective risk assessment without employing people who worked on voter protection issues for one or both parties and have local election administrators involved. I asked what they viewed as the highest risk parts of the infrastructure. He said there were states with no paper trail of how people voted and so focused there. I said I agreed the paperless machines were a big problem. But what about other potential risks, like hacking only enough machines that the decreased capacity would cause lines so long that tens or hundreds of thousands of people couldn't vote? Precluding 100,000 people from voting with long lines would be just as disruptive as changing 100,000 votes. It may be easier to get away with too. Russia would only have to hack machines so they broke and thus decreased voting capacity, not actually change votes. Broken machines could merely look like an unfortunate malfunction as opposed to a deliberate attack. As previously discussed in this book, unfortunately, people were used to long lines. It's unlikely anyone would even suspect malfeasance. At the time I didn't yet believe Russia wanted to get caught and as a result undermine confidence by demonstrating they can breach our elections.

This DHS official said creating long lines wasn't something they were even considering as within the scope of the risk assessment. We discussed other threats, for example, hacking the election results website, as the Russians did in Ukraine in 2014,[6] and as the Mueller Report later reported that the Russians were doing at that very moment to election websites in the United States.[7] Websites weren't something DHS was looking at either. Eventually it started to come out in the media[8] that Russia seemed to be scanning voter registration databases. I asked if they thought about the possibility of people being deleted from the voter registration database and lines being created from the ensuing confusion as people showed up at their usual polling place and were turned away because they weren't on the list. He said that wasn't being discussed either because they were told the people not on the list were able to vote on a provisional ballot.

I about jumped out of the booth I was so incensed by this point. I was like, who is giving you guys this information!?!? Yeah they can vote provisionally, but their votes aren't counted except in limited circumstances. While their vote could be counted in theory, in practice provisional ballots are almost never counted on the night of the election. So anyone asked to vote provisionally is, in practice, disenfranchised. As mentioned, John Kerry lost Ohio by only 120,000 votes when there were about 158,000 provisional ballots cast[9] and several-hour-long lines that cost him tens of thousands of votes. I had gone to absolute war in 2008, concerning the provisional ballot issue, to ensure our voters always cast a real ballot, never a provisional one.

Now that I was worked up, I was asking more pointed questions, like, "Well, when you say paper trail, do you mean paper ballot or something else?"

He meant that if there is a paper receipt that the voter gets after voting or that the machine stores, then that counts as a paper trail. I was astonished. I said, "First of all, the receipt that is kept in the machine is not something we can trust if the machine is hacked. There is no way to tell if it is storing the votes it is being told to store. That would literally require us to trust the record a hacked machine is keeping on itself."

What was even more disconcerting was that it seemed DHS considered the issues related to long lines and provisional ballots, and the website's unofficial results being hacked, more political in nature and not within the scope of what DHS was charged to do. DHS's role was solely to make sure the infrastructure functioned passably. When the election was over, DHS just wanted to ensure no one could claim the infrastructure was so compromised the election was invalid and would need to be rerun. Long lines, provisional ballots, and website issues were not things that would likely invalidate the election. Hacked machines with millions of votes deleted were more the concern of the day.

To him, the other issues seemed more like the types of issues the Clinton or Trump campaigns needed to be worried about, not DHS. He had a point. The federal government had never dealt with issues of long lines or provisional ballots. It had been left to the campaigns' voter protection programs in 2004, 2008, and 2012, to address issues related to long lines, provisional ballots, and so forth. Up to that point, those problems had been caused by U.S. election administrators who didn't know how to effectively run an election, not nation-states bent on undermining our democracy. It was reasonable for DHS to believe the respective campaigns would have robust efforts in place to ensure their votes were counted if basic administrative problems occurred, like long lines, regardless of what the feds were or were not doing.

My next meeting with DHS was even less reassuring. I was speaking with senior leadership at the ominously named National Programs and Protection Directorate (NPPD). Essentially, this was the cybersecurity hub of DHS. Unlike the colleague I had just met with on the analysis side of DHS, these folks were more outward-facing and dealt directly with those affected by cyberattacks. In this case they were dealing with state and local election administrators. Over a beer, I said, "So how is the election stuff going? Are you basically stuck doing that 24/7 these days?" His response shocked me.

> He said something along the lines of: No. We are staying as far away from this as we possibly can. It can only turn out bad for DHS and NPPD. We have no ability to fix anything before the election. No authority to force people to take our direction. No money, and it's not even clear we are "in charge" of the federal response.

A little stunned, I said, "Well, if not you guys, then who else at DHS or in the federal government would do this? It's not like TSA or FEMA can do this and the Bureau (FBI) doesn't really do prevention. The military can't do it. Not NSA. Who is left?"

"I don't know," he said. "But until someone tells us explicitly that it's our responsibility to fix we are not going anywhere near it."

This seemed like an odd response. NPPD always dealt with threats to critical infrastructure, and local governments were considered part of critical infrastructure. DHS had not yet called election infrastructure part of critical infrastructure explicitly, though later when the Russian meddling came into the public eye, DHS would come to designate it as such. There was always some nation-state hacking into our water systems or electrical grid or some other component of our critical infrastructure. This official was used to being called in after information surfaced that some breach occurred in another critical infrastructure sector. NPPD would get called in to work with the government and private sector players in question to assess what happened, work to remediate it, and put reports out for other critical infrastructure entities to patch systems for protection against the same attack.

NPPD always met some resistance when they showed up to an already miffed corporate or government entity that had just been hacked to essentially say, "I am from the government and I am here to help." This was part of a day's work for these guys. It was odd to me that he was so vigorously resisting this admittedly complicated but typical task for his team. It seemed these were no ordinary miffed local government officials.

He said I didn't understand. That no one wants their help. That they call and offer assistance but are told to stay out of that election official's business. He said they explain this is no ordinary threat, that this is a highly competent and motivated nation-state with essentially unlimited resources and clear intent to hack the elections, that it's not like it is some corrupt city councilman who may want to hack the election. This is possibly a team of highly skilled Russian hackers who may be doing this under the direction of Putin himself. He said the state and local election administrators still maintain they want no help, that those offering help don't understand elections. He said they don't get this type of immediate stiff-arm from other local government, electric grid, water system, finance industry, or other critical infrastructure leaders.

This was different. It was wholesale rejection of any security assistance that could be provided. To hammer home the point, this DHS official despondently said,

> Look, this is how bad it is. The other day I was on the phone with an election administrator, and after going around in circles trying to get him to accept the free cybersecurity assistance we were providing, he rejected everything I offered. Finally, I just said, "Okay, how about I just send you a list of industry

> standard cybersecurity best practices? You can look at those and decide for yourself which of them you want to implement." The guy said "no." He didn't want me to even e-mail him a list of best practices. I was taken aback. Of course, I asked why not? What harm could come from e-mailing him a list of best practices. He said that he assumed whatever the best practices were, he would not have the time or money to implement any of them and so didn't want there to be any record that he had received such a list. The election administrator said that if DHS was right and the Russians hacked his election system, he was worried about the subsequent fallout after the election. The official thought that if the Russian hackers did breach his system and alter the results, he didn't want there to be a record that he had ever received a list of best practices, because that could imply he was negligent in his duties and could be held culpable for the breach. He thought it better to be able to say that he never knew what cyber best practices were in the first place, so no one would say he is at fault for not implementing something he didn't even know existed.

I did a double take.

"Wait. What?"

The DHS official explained,

> The guy [leading election administrator] is worried he will be blamed for a successful attack. He doesn't want records showing he knew about measures he could have taken to stop the attack, since he figures the Russians could get around his defenses anyway. He wants to be able to claim ignorance and so not be seen as being asleep at the switch, when there were all these warnings about a possible attack.

"You have got to be kidding!?!?! That's gross negligence. A total abdication of duties. And, Jesus, talk about being more concerned with saving your own skin over defending your country. I mean, I wish I could say that it seems totally unbelievable, but based on what I dealt with in 2008, that sounds basically par for the course, albeit infuriating.

"Is that the reaction you are getting from all of them?" I asked.

"Some worse, some better. I mean the Georgia secretary of state is actively opposing us at every turn and has even accused us of hacking into his network."[10]

This was particularly frustrating for him, as my colleague at DHS was also a die-hard Republican and avid Trump supporter. He had zero interest in helping Hillary get elected and would be the first one to raise alarm bells if DHS even considered breaching a state network without their permission.

I was dumbfounded that he would be getting this much resistance from an election administrator. Not even wanting a list of best practices when there was a known threat looming was negligence of epic proportions.

I was disheartened at best when I left D.C. for the aforementioned Black Hat event in Vegas. But after the "Hackers for Hillary" fund-raiser there I was pumped up again and excited to offer help from the cybersecurity community in any way I could. So I e-mailed Robby once I was back from Black Hat about connecting with folks on his team to see if any of my cyber guys could be useful to him. Robby again registered real concern about Russian hacking of the election and their campaign and seemed genuinely interested in help wherever he could get it, in this case from the hacker community. So he connected me with people on his foreign policy and national security teams who would be dealing with Russian hacking issues.

This seemed odd. I thought I would be talking to the voter protection team. While the foreign policy and national security teams are made up of smart and capable people who thoroughly understand their subject areas, they generally would be the people on the campaign who know the *least* about how elections are administered in the United States. These are people who can spout out mountains of factoids about the Middle East, China, or Russia. However, they would struggle to name even one county clerk in the country or describe the difference in machines that count votes in Colorado versus Florida.

In contrast, it was the voter protection team's job to figure out how to ensure that every registered voter who wanted to vote for HRC got the chance to do so and had their vote counted accurately in the vote total.

Traditionally, the members of the national security team on a campaign are the wonks. Aside from giving speeches periodically, they have little to do with how a campaign is run. Most of their time is spent writing policy papers, preparing for debates, and helping the speech-writing team. The national security team is usually so removed from the real campaign team that they are often looked at askance by the campaign team, which views them as too snooty to get their hands dirty knocking doors or hanging banners at an event. Furthermore, as the campaign cannot direct the government, so the national security team cannot set government policy or direct federal agencies to take any action to stop the hacking anyway.

On the other hand, the voter protection team talks to secretaries of state and county clerks every day. They are positioned to ask, for example, if a particular election administrator has asked DHS for a list of cyber best practices to implement on their network. They can also demand to have demonstrations of what security practices are in place to protect voter registration databases. They could lobby to have independent cyber experts placed in key counties in battleground states so forensics can be done to identify breaches. Finally, the voter protection team works with the battleground state campaign team to generate media attention on local practices that are not up to snuff. They could call out local election administrators for not accepting help from DHS. These, among myriad other possible activities, are the kinds of things a voter protection operation could do. A national security policy team would not normally direct or carry out any of these tactics during a campaign.

I don't know how I wound up talking with the national security policy team. Who knows, maybe the voter protection team here or at the DNC was doing all the above things and more. In fairness to the campaign, hindsight is twenty-twenty, and no campaign in the history of democracy has ever had to think about responding to a nation-state executing a massive cyberattack on the infrastructure that administers its election. The campaign leadership was really making this up as it went along. There was no playbook or way to prepare for this in advance. It was unthinkable before 2016 that a country would try to hack our election infrastructure. It would not be surprising that the decisions they made weren't perfect.

Regardless, what stuck with me as I waited in the foyer of Clinton's Brooklyn HQ was what the first DHS official said. These issues of long lines, provisional ballots, and so on were up to the campaign to take care of, not DHS. DHS's role was to ensure the election infrastructure itself wasn't corrupted to the point that the election was invalidated. These issues about how elections could be poorly administered or tweaked by Russians to make poor administration worse, and thus disenfranchise more of a particular candidate's voters, was not central to the integrity of the overall process. Ensuring HRC's voters could vote was up to the HRC campaign.

When I finally was asked to go back and meet with the national security policy staffer, none of my concerns were allayed. She clearly didn't want to meet with me and was only doing it presumably because Robby asked her to. I mentioned the "Hackers for Hillary" fund-raiser and that I was happy to be helpful in any way I could. She said she didn't do fund-raising, in a way that could not have been more condescending. I said I understood that. I was simply highlighting that many of the people who attended the event might be useful to the campaign from a cybersecurity perspective. If they had a use for those cyber experts, I would happily organize a group that would likely volunteer for local campaign offices, or election offices or whatever the campaign wanted.

She responded by repeating such terse comments—dripping with condescension—as, "This was a national security issue," implying the hackers were not the national security establishment and thus irrelevant here. There were probably a few issues causing her to react in such a condescending way to me.

First, I was an Obama person, and she was from Hillaryland, dating all the way back to the State Department and the 2008 primary.[11] Second, I was a DHS person, and she was State Department. DHS was definitely viewed by the State Department and others as the ugly stepchild of the national security establishment, *especially* to the sophisticated State Department people. Finally, and this last one I was very sympathetic to, on a presidential campaign there is always an abundance of people who are offering "help." Usually that "help" is not help at all but a plot to enable the helper to promote some self-serving agenda, sometimes for policy reasons, sometimes for self-promotion, sometimes for money. There are a million reasons why people offer "help." As campaign staff, you never know what their angle is. I was told repeatedly this was particularly bad on the HRC campaign. The Clinton networks were so vast, there were always people finding a way into the senior staff to pitch some self-serving idea that was usually a bad idea to begin with and generally just led to more unnecessary work for the already overworked campaign staff.

For these reasons, and just because of her terrible demeanor, I could tell the meeting was going nowhere. I tried to wrap things up. As our short meeting ended, I made a throwaway comment that I would be happy to help the voter protection team with this issue if there was any interest. She then said something that still makes my stomach turn to think about today.

She repeated the Russian hacking was a national security issue. Therefore, they're letting the national security establishment deal with it. It was the national security establishment's mission to secure our elections. She had faith they could accomplish the mission.

Oh shit.

DHS had just told me they did not view it as their job to secure the election. Rather, they were just assessing things like which states had voting machines with a paper trail. Some election officials weren't even returning their calls and some would not even accept a list of cyber best practices over e-mail. Now this campaign staffer is saying she believed the national security establishment was securing the election so the campaign wasn't responsible for doing it. Oh shit. Were the federal government, state and local election officials, and the political parties all trapped in a complex "Who's on first" dilemma?

• 8 •

"We Have No Evidence"

DEF CON

After the election I bought into the argument everyone else did that the election infrastructure in the country is so disaggregated it would be almost impossible for the Russians, or anyone else for that matter, to successfully execute a nationwide cyberattack to change votes in a dozen battleground states. As mentioned previously, I do not believe Hillary lost Michigan, Wisconsin, or Pennsylvania because the Russians flipped votes. I believe she lost because she didn't campaign there, and her operations in those states were comparatively smaller than those of Democratic nominees in previous presidential years; however, the operative word here is "believe." I don't *know* this to be true; I am just basing my belief on some observations and previous knowledge. Really the only responsible conclusion one can arrive at is that they believe votes were changed or not changed. No one can say they "know" votes were or were not hacked because election officials didn't have cyber tools on their networks to identify such changes to vote totals.

So I was incensed when I began seeing all manner of election administrators on television and in front of Congress saying they had "no evidence" the election was stolen.[1] That statement is, I am quite confident, true. Still, I was plagued by a cognitive dissonance. My brain was constantly waiting for the next two sentences that must obviously follow the first. Those sentences are, to paraphrase,

> While we have no evidence of vote tampering, we had none of the tools or expertise in place at the time of the election to identify vote tampering in the first place. So to be clear, we have no idea if votes were changed or whether people were prevented from voting, and we will never know for sure.

In a perfect world, they would have followed up that statement with,

> Oh and by the way, we are completely incapable of defending ourselves from nation-state hackers. So if you all want to have any hope of preventing future attacks on our election infrastructure, our national security establishment needs to figure out how to defend our voting systems ASAP.

But, alas, in the wake of the Russian attacks on our elections in 2016, not one election administrator said anything remotely close to that. It appeared that for a few of the most vocal election officials, somehow the entire Russian attack on our democracy became about them. They were frantically trying to make sure they didn't look like they had been asleep at the switch during the election. In their mind this wasn't about the geopolitics of the Ukraine, Crimea, Syria, NATO, or EU expansion. Instead, Russia was a bit player in all this. This was a moment for a few vocal election administrators who never get time in the national spotlight to show they were competent. A steady hand in a time of crisis. Leaders even. Most of all, they wanted to make sure everyone knew that none of this was their fault. A convenient way to highlight this triumph of competency is to announce to the country, "Great news! We didn't find any evidence of vote tampering!" and just leave out the rest of the story that their office is so completely devoid of cybersecurity that no matter what happened in any election cyberattack, there could never be any evidence of vote tampering.

My concern wasn't that these election administrators should be taking the blame for having weak cyber defenses. No one should expect that they have the defenses in place to begin with. Even national security leaders thought the threat of another nation hacking our election infrastructure was infinitesimally low before 2016. Election tampering of another country is something the United States did to other countries like Guatemala decades ago. No one would have the gall to tamper with the United States elections. Or so they thought. The federal government, which funds national security, never gave election administrators a dime to secure their elections. In fact, this is playing out as it should. People on both sides of the political spectrum are blaming the national security establishment for not doing more to secure our elections. No one is blaming election administrators for not securing the elections, and rightly so.

What was driving me crazy—what was causing my cognitive dissonance—in the immediate weeks and months after the 2016 election, was no election administrator had the fortitude to say the other half of the sentence in a congressional hearing or on national television. They wouldn't say this is a key moment in our nation's history and the history of democracy. That this attack on our election infrastructure is a challenge of historic proportions. Our nation needs to rise to meet it with the full force the United States of America can bring to bear. Our national security apparatus should have begun opposing Russia at every turn. Silicon Valley should have been given a call to action to build the next generation of secure and transparent election equipment. Congress should have passed billions of dollars in funding to immediately upgrade our election infrastructure security. The State Department should have begun convening our allies immediately to ensure we work together to protect their elections, as well as those of emerging democracies. Obviously, none of this happened. It was largely because the national security establishment knew almost nothing about how elections were administered. The election administrators reassured them everything was fine and Trump didn't want to talk about Russian interference in the elections at all. So no whole of government response was coming. So it was that at a March 2017 conference of progressive leaders, technologists, and media experts in San Francisco that I had finally had enough.

I was leading a discussion on cybersecurity at the conference and noticed no one was talking about election infrastructure. Everyone was talking about fake news or protecting their campaigns' e-mails and texts from being hacked, à la John Podesta. I, half-joking, made a side comment that touchscreen voting machines are the work of the devil. A couple people pulled me aside afterward and said they also wished there was more of a national debate about how to secure our election infrastructure. We agreed the fake news stuff was a problem, but the integrity of our election infrastructure was a far graver threat. And now that Trump was president it seemed like little to nothing would be done on this because the issue was dying in terms of public interest.

I began thinking about what could be done to ensure policymakers and the public were aware that what these vocal few election administrators were saying was disingenuous. They were putting their own narrow personal insecurities ahead of our national security. Our election systems were a sieve. Terms they were using to describe our election infrastructure to Washington and the media—"air-gapped" and "unhackable"[2]—were not only inaccurate, but also dangerously irresponsible, causing a sense of security that played into Putin's hand and made our democracy more vulnerable.

After the conference I went out for dinner and drinks with some colleagues who were experts in politics, tech, and media. I lamented we would all be lulled into a state of complacency. At about midnight, I ambled back to my hotel room in the Mission District of San Francisco, frustrated and feeling wholly inadequate. I knew this issue as well as anyone in the country. I was with some of the best minds in exactly the right industries to help me find the best way forward. Yet, I had nothing. No idea what to do. I was just a typical whining, impotent progressive, complaining about the state of affairs in the country with no good idea about how to fix it.

In the morning I headed out to get some coffee and see the last couple sessions of the conference. As I walked out of my hotel, I was looking at my phone for directions and next to me was a pale, stocky homeless man muttering something to himself and pacing back and forth in small, three-step laps. Then another homeless man, tall, pasty white, tatted up with an undershirt on, walked about two feet in front of me and bashed the other guy in the face with a coffee cup. The other guy snapped out of his trance and stood in the middle of the street and ripped off his shirt, growling and then throwing punches as the two erupted in a massive brawl in the middle of the street at 8 a.m.

I stepped out of the way and began walking toward the conference. About ten steps away, as I kept walking from the street fight, I saw a homeless woman (I think it was a woman), leaning up against a wall with the back of her skirt hiked up. She was taking a massive shit on the sidewalk. I put my earbuds in and kept walking. During my entire jaunt to the conference, I was thinking what an insane 45 seconds that had been. I felt like a character in *Blade Runner.* I was headed to a high-tech conference with the political elite in the country while pandemonium among the homeless reigned outside. It was both hilarious and sad, but most of all jarring. The rest of the day I was in a little bit of a haze, my thoughts bouncing between the conversation on election hacking from the previous night and the fight/shit this morning.

Sometime between the fight/shit and leaving the conference for the airport, an idea popped into my head. It didn't evolve during the course of the conference, nor did I construct it by combining several ideas I had been chewing on for some time. It just popped into my head. I thought, "Huh. That's interesting." I immediately broke out of the haze caused by the crazy events of the morning.

In the cab on the way to SFO, I texted Jeff Moss, who runs DEF CON, described earlier. I wrote, "Hey, gonna call you in 20 min when I am through security. Pick up." Shockingly, he responded immediately. He is notoriously hard to contact and lives in Singapore, so I almost never get him to respond immediately.

After I got through security, I called him and said, "Hey man, know how you guys got all that coverage at DEF CON last year when you all hacked into cars?"

"Yeah, that was awesome," he said.

I replied, "Well with everything going on in the news since the election, why don't you guys hack into voting machines at DEF CON this year?"

There was a long pause, and then he said, "Fuck. That's a good idea. Let me call you back." And he hung up.

We traded texts a few times, and I finally got him back on the phone about a month later. Unbeknownst to me, after we hung up, he immediately went on eBay and bought every electronic voting machine he could get his hands on. Then he called Harri (the hacker from chapter 6) and a professor from the University of Pennsylvania named Matt Blaze, and asked them if they could manage the technical aspects of the "Vote Hacking Village" at DEF CON, which was being held in July at Caesar's Palace in Las Vegas.

When Jeff and I spoke again, we began to game out what we would do. Fortunately, eBay had tons of voting machines that were the same make and model of those used throughout the United States today. Moss and I then leaked to Politico that we would be hacking into the machines at DEF CON. Eventually people began coming out of the woodwork to offer Moss, Blaze, Harri, and me voting machines. It was not lost on us that the election administrators' claims the Russians would not be able to hack our machines because they had no prior knowledge of the machines, and therefore it would take them too long to hack without being noticed,[3] failed to pass any sort of reality check. Any Russian intelligence officer who wanted his team of hackers to have machines to practice on could just go on eBay and buy as many as he wanted.

In fact, our problem was not getting machines, but being able to buy just one or two of a certain type instead of pallets of hundreds of machines at one time. The Russians could give each member of the entire GRU (Glavnoye Razvedyvatel'noye Upravleniye) their own machine to practice on at home. Also, the election administrators *must* have known these machines were available on eBay for anyone to buy and practice hacking into because *they were the ones selling them.*

It turns out that when a jurisdiction gets new machines, or if they have overstock, or if the exterior casing is damaged, election administrators routinely sell the machines to other jurisdictions. They often sell them on eBay or other sites. There is certainly nothing wrong with this practice. It is perfectly legal and ethical; however, the problem arises when election administrators would yet again lull the public into a false sense of security by implying the machines were being guarded at Fort Knox in the downtime between elections.

The other disconcerting aspect of the availability of voting machines for Russians to practice hacking into—which we didn't realize until after the conference—is that the software on the machines didn't upgrade throughout time to patch bugs. Meaning if the machine was manufactured in 2003, it may still have had 2003 software on it in 2016. This was not the manufacturer's fault or that of the election administrators. It was the fault of the federal Election Assistance Commission (EAC), whose certification process is so arcane and glacial in pace that vendors and officials could not update software and remain compliant with the EAC's certification standards.[4]

This was also something the election administrators absolutely knew while the vocal few were telling the public that the Russians would need significant prior knowledge to hack the heavily guarded machines successfully when they were briefly released from protection on Election Day. The machines ran on a host of commercially available software, for instance, Windows XP, that was more than a decade old. I think it is safe to say that Russian hackers know how to hack Windows. In fact, anyone with access to Google can hack Windows XP because the tech community has been publishing information about the exploits in Windows for years. This is the equivalent of peer-reviewed research in academia but for hackers. Hackers figure out how to break into a certain piece of software, and then they put their exploit online to both show off to their friends and so their colleagues can check their work. They also do it so the software company can fix the bug. Microsoft welcomes this type of research and actually pays hackers for the bugs they find, as long as the hackers tell Microsoft before publishing it so the manufacturer can fix the problem in the next "Patch Tuesday," which is the day Microsoft sends updates to your computer to fix itself.

That is, Microsoft sends updates to your computer, but not your voting machine, because the EAC doesn't allow these automatic updates to be made.[5] Vendors and election administrators had been complaining about this to the EAC for years. So there is no way election administrators spouting the virtues of the physical security of these voting machines didn't know that neither were the machines heavily guarded, nor was much of the software in the machines unique.[6] The software didn't even possess the most rudimentary, publicly available system updates.

Voters had no way of knowing any of this until the hackers at DEF CON ripped open the machines for the first time publicly to find out what was inside. We started checking around after we decided to organize the Vote Hacking Village at DEF CON and found out that there had *never* been a public assessment of the security of the machines quite like the one we were proposing to do at DEF CON. We were astonished to find that in the several decades that these machines had been in use, only a handful of security assessments had been done by disinterested third parties. Moreover, these assessments were always done under strict nondisclosure agreements, where researchers essentially had to sign away their first-born child if any details got out of what exploits were found.

Based on my experience in 2004, 2008, and 2016, it should not have been surprising to me that there were no public assessments done of the voting technology; however, I was still amazed.

Moss was also amazed they hadn't hacked voting machines at DEF CON before. He checked and said speakers at the conference first raised concerns about security of voting machines more than a decade ago.[7] They had hacked into about everything else at this point: medical equipment like pacemakers, several cars, all manner of social media sites, and almost every type of software you use on your phone or computer. They even had a section one year on hacking adult toys. Voting machines may have been the only thing they hadn't hacked.

Next we decided the hackers at the conference, the media attending the conference, and the policymakers I wanted to influence later, needed context about why any of this mattered. There was still a strong push by the election administration community to make response to the 2016 attacks a bureaucratic voting administration issue instead of the national security issue it clearly was. So our first step was to lock in several keynote speakers for DEF CON who hailed from the national security world. We also wanted to ensure we had an understanding from DHS and other agencies about what we were doing. So I began privately briefing key cyber experts at DHS about our plans in advance of DEF CON.

I also started to reach out to several election administrators and the EAC about what we were planning. I repeatedly made the point that we had no interest in making any election administrators look bad in this exercise. I made the point that the hacker community was the last group that would blame the election administrators for Russian hackers getting into their systems. My common refrain was that we weren't attacking the election administrators or vendors, or anyone else for that matter. We were attacking the *idea* of unhackability with respect to elections.

I was pleasantly surprised when many of the election administrators began saying they were cautiously optimistic about what we were doing. Some even said they were happy we were doing it because the voting machine vendors had been telling them for years their machines were "unhackable." The officials said they never believed it but didn't have a way to disprove them. Others told us they planned to send their staff to DEF CON to learn what the hackers found. So they could better understand vulnerabilities, which was music to my ears. It seemed like maybe we were seeing a sea change in election officials' attitudes toward voting security, or maybe this was the silent majority who weren't proclaiming to the press and Capitol Hill how secure our elections are.

The then head of the EAC, Matthew Masterson, even suggested we disseminate the technical resources on their website so the hackers could prep their hacks in advance. He also sent his most technical senior staff, Ryan Macias, to attend and learn from the event. Again, we were pleased to see the sophisticated response the EAC leadership had to our plans.

In this process, I also got to meet Noah Praetz, who was director of elections for the Cook County clerk, David Orr, in Illinois. Noah was light-years ahead of most others in this space. He saw this problem for what it was: an existential threat to the United States from a powerful foreign adversary. He was intrigued by what we were doing. So he gave us the specs to his own county network. With those specs, we recreated a realistic clerk's network on a virtual cyber "range" where election IT staff could train at DEF CON on how to defend a recreation of a real clerk's network against the best hackers in the world.

As the conference neared, several activists in the voting security community began getting cold feet. They had people call me and the other organizers and say, "You shouldn't do this! What if the hackers can't get into the machines?!? It will reinforce the vendor claims that the machines are unhackable!" I guess in some of the previous studies done privately, researchers were given a full week and all the source code and still had a difficult time breaking into the machines.

Other activists worried that, "Well, the vendors will just say they have more updated equipment than what you are hacking so what you are doing is meaningless."

Similarly, "The vendors and election administrators will say that this exercise isn't fair because it's not being done in a controlled environment like the previous private tests."

Another ran something like "If you highlight these vulnerabilities for the public, you are just playing into Putin's hands. You will convince people our election infrastructure is so insecure people will stop trusting election results. Eventually they will just stop voting altogether."

I routinely had to answer these concerns with something to the effect of,

1. "The Russians didn't play 'fair' and definitely would not be hacking these machines on the vendors' and election officials' terms."
2. If the hackers at DEF CON did breach the machines, the press, policymakers, and the public would not for a second buy that whatever slightly better software version the vendors had was unhackable. . . . Moreover, while the vendors had newer equipment, the machines we had were by-and-large being used around the country today.
3. I had total confidence these hackers were some of the most clever people on the planet. They would get into the machines before the weekend was out.

4. This is the attack I was most concerned about. I had spent a large part of my career trying to increase the number of voters. I didn't want to do anything to dissuade people from voting. This argument usually came from the vocal few in the election industry and sounds deceptively logical. So I will go into a little more depth as to why this is both theoretically and practically wrong. I refer to their argument as the Chinese Communist Party (CCP) Argument. The comparison is apt because it suggests dealing with the problem exactly how the CCP deals with problems in their government. When a problem is identified, the CCP immediately tells everyone in the country they are not allowed to talk about it because it will just make the problem worse. In effect, they say, "Trust us. Now that we know about the problem, we will quietly fix it behind closed doors. Getting the public worked up about this will only slow our efforts to fix things and cause unrest in the process. Also, its unpatriotic to criticize the CCP. You are naively playing into the American's hands." In over 200 years of existence, I think America has found a better way. Instead, when our society finds problems with government it shines a light on those problems and attacks them from all angles. We talk incessantly about our problems with government until we fix those problems and usually well after. As an American and former government official who has been on both the criticism giving and receiving side of the equation, I think this is a far better formula for success. Moreover, the fact remains that our systems are insecure and we will be attacked again. So even if you play out their CCP argument, it's still the wrong way to go. Let's say we tell the public election systems are secure and then they are hacked again. We spend years assuring them everything is "air-gapped" and "unhackable" and then our elections are hacked. No one would trust local or federal officials again. Being caught in a lie or obfuscation of the truth is the best way to lose the public's trust in government. On the other hand, if we acknowledge the security flaws. Remind the public this is a categorically new problem we had not encountered before 2016. Then demonstrate publically we are transparently doing everything we can to find all the flaws we don't know about. Subsequently we will fix those flaws. Remind the public that we can not keep all the bad guys out all the time. However, we are going to make it a lot harder for them. Further, we will increase our chances of detecting successful attacks by implementing good cyber hygiene, paper ballots, and audits. Then, when we are inevitably hacked again, the public will know we have been on the case. They won't think we lied to them and will be receptive when we come

> back with a transparent plan on how we will become even more secure. Finally, we didn't know it yet but turnout for the 2017 and 2018 elections was up sharply from previous elections. So we have millions of data points in the form of real votes, that demonstrate all the publicity around hacking voting machines and Putin has not decreased turnout. So it turns out the CCP argument is not only theoretically but also practically wrong.

As DEF CON neared, we were getting it from all sides both good and bad. Some of the bed-wetting activists who got cold feet as the conference approached were blowing us up every day with their concerns about it backfiring. Conversely, we had some incredibly positive interactions planning DEF CON with DHS, the EAC, and more sophisticated clerks like Noah in Cook County, Illinois; Ricky Hatch in Weber County, Utah; Dean Logan in Los Angeles County; Barb Byrum in Ingham County, Michigan; John Odom in Montpelier, Vermont; and Joel Miller in Linn County, Iowa. However, a few secretaries of state who were closely affiliated with their D.C. advocacy group, the National Association of Secretaries of State (NASS), really began to freak out in the days before the conference and made one last-ditch effort to shut us down.

They were smart in that they tried to shut down Moss and me by going after something they knew we both valued deeply: our relationship with DHS. I was still on contract with DHS as a consultant for cyber policy issues. This was a tenuous position, as I was a former Obama official and we were six months into the Trump administration. Moss was on the Homeland Security Advisory Board to the secretary of Homeland Security (then General John Kelly) and being considered for renewal.

We had set up several meetings for the soon-to-be deputy assistant secretary of the Cyber Division of DHS, Rick Driggers, in Vegas at DEF CON with some of the key individuals involved in the Voting Village. Rick was sincerely interested in what these top hackers had to say about their experience in the voting security arena. The election security portfolio was handed to him and, like everyone else in the national security world, he was trying to get smart on securing elections. He was asking these top hackers such nuanced questions as, "How do you think we should define where a voting system begins and ends?" "What am I actually trying to secure?" was a question he asked repeatedly in the days leading up to DEF CON.

On the eve of DEF CON, which was set to take place on July 27 through July 30, 2017, a small group of state election officials and NASS staffs, who were also the most vocal in claiming our election infrastructure was "air-gapped," "unhackable," and that there was "no evidence votes had been changed," got together. They were at a two-day gathering in New York with a group of DHS senior officials talking about what was to be done to safeguard election security. These conversations in the early days generally had little to do with what these secretaries were going to do to secure their networks, but rather about telling DHS what it could *not* do because the locals were "in charge" of elections. Some discussions revolved around activities like getting all the secretaries of state security clearances. By the way, almost none of their counterpart agency heads have security clearance despite the fact that they are all being hacked by Russians and others constantly. For example the state head of the department of transportation or state labor department get hacked all the time. However, DHS has yet to give most of them security clearances. These other state agency heads don't have security clearances because they don't need to know what division of the Russian GRU is hacking them. They simply need to listen to the cyber experts at DHS and spend their limited time and resources implementing basic cybersecurity hygiene, paper ballots, and risk-limiting audits. No one needs a security clearance for that. Unsurprisingly, I have noticed there is an inverse relationship between how often a secretary of state mentions they have a top-secret security clearance and how much they have done to actually secure their networks.

While a few of the secretaries were complaining to DHS about their litany of grievances, they said something along the lines of the DEF CON Voting Machine Hacking program was a total "stunt." They requested DHS direct its senior officials (namely Rick) at the event tell Moss and me that he would use his portion of the speaking program to say that the entire thing was a disingenuous stunt. He should go on to say neither he nor DHS would have any part of it and then walk out.

These few vocal secretaries of state and the NASS staff clearly were worked up. They scared a DHS official at the gathering so much he called Rick in Vegas and told him DHS leadership expected him to follow the direction of the secretaries. Fortunately, Rick had just finished a rational and eye-opening dinner with Harri and Blaze when he got the call.

I was totally taken aback. I felt like I had just been punched in the gut. For months I had been working tirelessly to keep the senior-most cyber officials at DHS, EAC, Congress, and dozens of state and local officials in the loop on what we were doing. I had solicited the input of DHS's hackers on how to set up the Village. I had done everything you are supposed to do when trying to get buy-in from a huge agency like DHS. But now a few parochial, insecure luddites were going to use their political leverage over DHS officials to kill the whole thing. I was really worried that someone from DHS was having the same conversation with Moss that they just had with Rick and that Moss may be inclined to shut it all down. All this was rushing through my head as I tried to read the expression on Rick's face.

Then he said something like, "No way I am doing that. I get what you guys are trying to accomplish."

Whew.

I want to pause for a moment and highlight one of the few bright spots in this story. Rick was about to be nominated as a presidential appointee in the Trump administration. He and his two bosses at DHS work for a president who thinks the election hacking is a hoax. The topic is so radioactive that the secretary of homeland security isn't even allowed to bring it up with him.[8] Anyone who gives credence to the Russian hacking claims runs the risk of being tubed by the White House. Despite this, many of the senior most cybersecurity officials at DHS like Rick Driggers, John Felker, Rob Karas, and Doug Maughan have pushed the envelope time and again, getting very far out over their skis with absolutely zero top cover from the White House. Rather, they were working with an enormous amount of risk that they could be fired for focusing on this. But they do it anyway. They worked tirelessly to get election administrators the resources they needed despite some recalcitrant election administrators who were unappreciative of the great professional risk officials like these four individuals were assuming. Despite being in a department with plenty of Trump die-hards who would be more than happy to make their lives miserable, these officials pressed on and quietly made this a top priority for the department. By the way, most of those I mentioned above are Republicans (I won't identify which are GOP and which are not to protect their privacy). They have no interest in working on this issue to somehow make the president look bad. These officials are a testament to the fact that there are still real patriots—great Americans—at the Department of Homeland Security who put country over party and self. Even though they are given a shit sandwich every thankless day on this topic, they drive it forward because they know it's the right thing to do.

In addition to reinforcing that Rick is a great guy, I began to get an inkling that we were on to something bigger than anticipated. The secretaries of state and NASS staff got so worked up they threw down a dramatic ultimatum. The activists got so worked up they were haranguing me constantly. It was clear we were striking a chord. People were paying attention.

Further evidence this was striking a chord was the deluge of media inquiries we got after announcing the full scope of what we were doing. In the days leading up to the event, Moss, Harri, and I were each doing several interviews a day. Every major media outlet in the country and many others throughout the world wanted to cover it. We even had members of Congress who wanted to come observe the event.

When I woke up in Las Vegas on the morning of DEF CON, we had election officials throughout the country gunning for us to fail. A bunch of well-connected and loud-mouthed activists were wrapped in knots waiting for us to fail. So they could publicly say "I told you so." At the same time, some brave local election administrators, like Noah, and senior DHS leadership, like Rick, were going out on a limb to defend us with their personal credibility. And just because I am a glutton for punishment, I decided to bring my wife and six-year-old twin children along to see everything. So it was with all this in mind as I stared down at my kids looking at me bright-eyed as I was leaving the hotel room and asking, "Are you going to hack the votes today, Daddy?" I fucking hope so.

As I opened the Voting Machine Hacking speaking track, Harri and Blaze opened the Voting Machine Hacking Village. They stood at the front of a large ballroom filled with long tables piled high with electronic voting equipment, electronic pollbooks, memory devices, cords, and other computer paraphernalia. There were more than 25 pieces of election equipment, some of which included an AVS WINVote DRE, a Premier AccuVote TSx DRE, an ES&S iVotronic DRE, and a Diebold ExpressPoll 5000 electronic pollbook. Most of the equipment had never been tested in such a public, unrestricted setting. They did about an hour of intros, explaining what the equipment was used for and so on. Then they just stepped back and let the hackers have at it.

Within minutes Harri came running over to the speakers track room where I was getting ready to introduce our next speaker. He grabbed me and said something like, "They already hacked the first machine. This Danish guy hacked it remotely two minutes after we gave them access to the machine." His eyes were as big as saucers. I was like, "Holy fuck? Really? You guys got in already?"

I grabbed the mic from the front of the room and said something like, "Excuse me, excuse me, I would like to make an announcement. The first machine has been breached. So within minutes of having access, one of the hackers, with no prior experience hacking voting machines, took over the machine remotely."[9]

The press in the room went nuts. Hands immediately began shooting up in the air. There were lots of questions. I just looked at Harri and said something like, "Fuck, dude. Game on."

My original fear was the Voting Village results would be underwhelming and give ammunition to the naysayers. However, with a machine hacked remotely in the first two minutes of the conference, I then found myself wondering how we might fill the next two days. I was hoping we hadn't hit the grand finale on day one, hour one. Turns out, that fear was unfounded. The following are excerpts from the report our team wrote following DEF CON in 2017.

The results were sobering. By the end of the conference, every piece of equipment in the Voting Village had been effectively breached in some manner. Participants with little prior knowledge and only limited tools and resources were quite capable of undermining the confidentiality, integrity, and availability of these systems. Moreover, a closer physical examination of the machines found, as expected, multiple cases of foreign-manufactured internal parts (e.g., hardware developed in China), highlighting the serious possibility of supply chain vulnerabilities. This discovery meant that a hacker's point of entry into an entire make or model of a voting machine could be well before that voting machine rolls off the production line. With an ability to infiltrate voting infrastructure at any point in the supply chain, the ability to synchronize and inflict large-scale damage becomes a real possibility. Also, as expected, many of these systems had extensive use of binary software for subcomponents that could completely control the behavior of the system and information flow, highlighting the need for greater use of trusted computing elements to limit the effect of malicious software. In other words, a nation-state actor with resources, expertise, and motive—like Russia—could exploit these supply chain security flaws to plant malware in every machine and, indeed, breach vast segments of U.S. election infrastructure remotely, all at once.

The first voting machine to fall—an AVS WINVote model—was hacked and taken control of remotely in a matter of minutes using a vulnerability from 2003, meaning that for the entire time this machine was used from 2003 to 2014, it could have been completely controlled remotely, allowing the changing of votes, observing who voters voted for, and shutting down the system or otherwise incapacitating it. That same machine was found to have an unchangeable, universal default password—found with a simple Google search—of "admin" and "abcde." [In fact, hackers were able to "Rick-Roll" the machine, causing Rick Astley's 1987 hit "Never Gonna Give You Up" to blast from the voting machine's speakers.]

Physical access to the machines afforded similar ease of access. The locked panel on the front of some devices was easily picked or opened with readily available keys that could be purchased easily and cheaply online. The locked cover was easily bypassed without paying attention to the lock as well (by simply compromising the plastic hinge). The physical security protecting the USB port was ineffective and also irrelevant due to the findings in the following paragraph.

For other machines, all the hacker had to do was simply attach the keyboard and type "ctrl-alt-del," and the Windows task manager would pop up. At that point, they could type "alt-f run" and run any software they wanted to, including software on a USB stick that is inserted into the other USB slot on the back of the machine. As one hacker, Nick, remarked, "With physical access to back of the machine for fifteen seconds, an attacker can do anything."

·9·

"Child's Play"

DEF CON exposed the insecurities of our election infrastructure. We were determined to ensure that policymakers and the public were aware of these vulnerabilities. Our activities at DEF CON had been widely covered by the press, but that coverage had not penetrated the media bubble that the nation's national security experts live in. These are the people who have copies of the prestigious journal *Foreign Affairs* on their nightstand instead of *Wired*. And these experts were our top targets in advance of a public policy and legislative push.

We decided to issue a report on the voting machine vulnerabilities identified at DEF CON. The report was going to be released in Washington. Doing so would make this a political issue, but we did not want it to become a partisan issue. In an effort to show the nonpartisan nature of our work we partnered with the Atlantic Council, a center-right think tank in D.C. headed by Fred Kempe, the former CEO of the *Wall Street Journal*, and released the report in the Brent Scowcroft Center at the Atlantic Council. Brent Scowcroft, former national security advisor to President George H. W. Bush, is one of my heroes so releasing the most important publication I had ever worked on in a prestigious center bearing his name was particularly appealing to me. With the Atlantic Council as a partner, we took our first foray into cyber politics in Washington.

Despite the flashy headlines at DEF CON of machines remotely hacked in two minutes, we decided to highlight a more mundane problem with voting equipment that was obvious but previously unreported and far more consequential. After DEF CON, Harri texted me a picture of the back of a voting machine he had opened at DEF CON. The parts read like a testament to the modern global supply chain with parts labeled, "Made in China" or "Made in Taiwan" but nowhere in the picture was a label reading "Made in America." Harri sarcastically wrote something like, "The stamp of approval for rock solid security." This was a real security flaw and I wanted to highlight it in the report.

I argued with the other technologists helping with the report for weeks about the significance of our voting machines essentially being "Made in China." Most of the technologists argued either that it was unfair for us to point out this concern because "everything else is made in China" or that the voting machine vendors didn't have much choice but to have the machines made in China because they could not meet the low-price demands of the election administrators if they didn't make their products in China or another country with low-wage workers.

I was having none of this. Obviously, the points the technologists made were true. However, it was also true that Russian hackers had a long history of hacking into electronics factories in China (and elsewhere) to infect chips destined for machines in the United States. In 2018, Bloomberg News reported on a supply chain hack originating in a Chinese factory. Chips had been inserted into hardware by the People's Liberation Army, and then sent out to 30 companies, two of which were reported to include Apple and Amazon.[1] Both companies vehemently denied this claim.[2]

We also know that both FedEx and Maersk lost about $300 million each from a tweaked version of M.E.Doc, which allowed malware to be delivered in the guise of ransomware. Ultimately, the attack was attributed to Russia.[3]

We were releasing our report to national security experts who are deeply versed in the global supply chain threat posed by Russia and China. We still needed to convince election officials and the public that the supply chain threat is not a dystopian nightmare. Rather, it is a real threat that Russian hackers have exploited often in other industries.

Moreover this threat strikes at the heart of the vendor and election administrators' argument that remote attacks on election infrastructure are impossible. As noted earlier, they claim the elections are run by so many tiny jurisdictions it would be almost impossible to scale a remote hack affecting millions of votes. While there are 8,800 election jurisdictions throughout the country, there are only three major vendors. If the Russians could infiltrate a chip manufacturing plant in China or elsewhere, they could hack entire classes of machines throughout the country, all at once, from the Kremlin.

Furthermore, we noted in the report that while some progress was being made in securing statewide voter databases, virtually no progress was made securing voter registration databases kept at the county level where elections are actually administered. At the report's release, we made these points and others listed in the last chapter. We also had heavy hitters such as the former U.S. Ambassador to NATO, General Douglas Lute, as well as the only human ever to run both offense and defense at NSA, Sherri Ramsay, there to back us up. They explained the gravity of the threat and Russia's determination to undermine the fabric of our society from within.

Nonetheless, despite these well-researched positions and the stable of some of the most reputable leaders in the national security establishment endorsing our claims, the top staff from the National Association Secretaries of State (NASS) and some of her colleagues were there heckling us during the event. They were demanding to know why there were "all these generals and cyber people up there" instead of election administrators. This complaint ignored the fact that John Gilligan, head of the Multi-State Information Sharing and Analytics Center (MS-ISAC), was one of the speakers at the event. His organization is funded by DHS to provide cybersecurity for state and local governments. Further, the MS-ISAC had just been tapped *by the secretaries of state* to be the facilitator for state and local election security.

NASS is a recurring character in this story despite almost no Americans knowing who they are or what they do. Thus it may be prudent to take a moment and examine their role in this saga; the National Association of Secretaries of States' stated reason for being is to "serve as a medium for the exchange of information between states and [foster] cooperation in the development of public policy."[4] This mandate includes valuable activities for our country like educating federal officials on the needs and challenges facing state election officials. NASS also does great work convening conferences where election industry training is abundant. Additionally, there are many fantastic individual secretaries of state who are part of the national association like Alex Padilla in California, Mac Warner in West Virginia, Denise Merrill in Connecticut, and Kim Wyman in Washington, just to name a few. These secretaries of state are taking advantage of all available resources to improve the security of the election systems they oversee. They, and many others like them, are true public servants. So, to be clear, NASS is a legitimate organization with an important mandate that does some very valuable work.

That being said, NASS's role in federal policymaking is far less benign. As we will see here and later in the story, they have been active in undermining any attempt to point out election infrastructure security flaws to the public or policymakers. They were instrumental in killing our best chance so far at common sense election security standards or legislation.

The staff at NASS act as the de facto top federal lobbyists for the secretaries of state in Washington. That in and of itself is not nefarious. Like many other state and local government entities, the secretaries need representation in Washington. The problem is that, because of typical Washington loopholes and lack of transparency that drive the American people crazy, NASS staff doesn't have to register as lobbyists, so journalists and other watchdog organizations have no way to shine light on how this largely unknown organization is dramatically affecting federal policy around elections. For example, NASS wanted a no-strings-attached check to go to each state.[5] It won that battle in Congress and at the federal Election Assistance Commission (EAC). Now states can spend federal grant dollars meant for security on basically whatever they want. NASS also killed the leading bill in the Senate for common sense, bipartisan legislation mandating election security standards. Since the attacks in 2016, NASS's entire Washington operation has constantly bent to the wishes of a few members who expect the staff to kill any measure designed to mandate higher federal security standards for election infrastructure. In our case, it seemed that they also tried to stymie any attempt to shed light on the real security threats that exist in our elections. This trifecta of killing restrictions on how federal election security funds can be used, killing bipartisan election security reform bills, and attacking our attempts to highlight vulnerabilities is less than benign at best.

With all these nefarious actions coming from an otherwise useful organization, one must ask "Why?" Why does NASS deliberately torpedo legislation intended to give our nation's election administrators the resources and tools they need to do their job? Especially when, as stated repeatedly above, many of their members want to implement robust security and many of their members want federal funding to do their job better. We may never know all the inner workings of NASS, but two things are certain, as they have been said to me directly from members of NASS and members of the Congressional committees they lobby.

First, NASS is often held hostage by the "Tragedy of the Commons" trap of gaining consensus among a group of fifty secretaries of state that vary drastically on the political spectrum. Many of the secretaries I have mentioned above and elsewhere would gladly agree to some common sense federal security requirements in exchange for receiving desperately needed federal funds. They view this as akin to their counterpart state secretaries of transportation agreeing to federally mandated speed and alcohol consumption limits in exchange for federal highway funds. A no-brainer corollary would be accepting federally mandated voter marked paper ballots, audits, and cloud-based voter registration databases in exchange for federal funds to implement those practices. However, some secretaries make a "state's rights" argument that no federal requirements are acceptable because they would infringe on state's rights. This is obviously ridiculous because the federal government already mandates some election standards under the powers assigned to Congress by the U.S. Constitution. Further, the U.S. Constitution is abundantly clear on who is assigned the task of protecting our democracy from foreign nation-states. It is not the secretaries of state but rather the federal government. Despite the U.S. national security apparatus' obvious responsibility to protect our democracy, NASS can generally only achieve consensus among its members to lobby for (1) federal funding with no strings attached and (2) no new election security legislation mandating even basic security practices. Of course this is despite the fact that most election officials have no expertise in cybersecurity or geopolitics. That's not their fault. They didn't get their job for expertise in cybersecurity or geopolitics. Again, what is NASS's fault is torpedoing *free* help and standards from those who do understand cybersecurity and geopolitics.

The second eyebrow-raising issue as reported in *The New Yorker*, NASS receives an amount estimated to be in excess of six figures annually from election industry vendors in the guise of a "corporate affiliate program."[6] These election vendors sell neither cybersecurity nor election audits. They benefit from no-strings-attached money going to states because that enables the states to buy products they do sell rather than cybersecurity or election audits. So it is in the vendors' interest for NASS to lobby, under the auspices of speaking for the secretaries of state, for states to receive election security funding without restrictions on how that money can be spent.

Further, through a series of political organizations too complicated to explain here, some of the vendors fund the secretaries of state political campaigns to the tune of tens of thousands of dollars. It should come as no surprise that the secretaries who are the staunchest supporters of "no-strings-attached funding" and "no new security standards" policies are the ones who receive the most money from the vendor funded PAC. Whatever the driving factors, the end result is the same: NASS has consistently lobbied against common sense election security requirements being tied to any federal funds for states. We don't know whether vendors are demanding secretaries receiving campaign dollars advocate for the vendors' positions within NASS or whether NASS staff receiving sponsorship dollars are expected to do the same. However, it is an eyebrow-raising coincidence to say the least.

We all pay the price for NASS's undisclosed lobbying efforts. No-strings-attached funding and no new security standards are obviously nonstarters for most Congressional members; so NASS's lobbying has had the unfortunate side effect of ensuring no new funding has been appropriated for election security (the $383 million appropriated was old HAVA money) and no new security standards have been mandated since the attacks of 2016.

We can't necessarily fault NASS for being held captive by the lowest common denominator policies that consensus ensures. The organization does good work in other areas and it would likely not be able to exist if it didn't have this "consensus" rule in place. But just because the process by which they arrived at their policy stance isn't nefarious doesn't mean the policies they are lobbying for aren't nefarious all the same.

While they don't have to register as lobbyists, they are advocating positions to the federal government as a lobbyist does. Thus they exist in an opaque pseudo-lobbyist world where their actions are neither disclosed nor tracked. On top of this, as reported in the *New Yorker,* NASS representatives portray themselves solely as speaking on behalf of fifty secretaries of state when speaking to members of Congress and their staff. The congressional staff reported that NASS does not actively disclose the gigantic sums they reap every year from the election equipment vendors when in conversations with members of Congress. To be clear, it is neither illegal nor immoral to take money from industry, and NASS lists its "corporate affiliates" on its website. But once an organization is accepting significant financial support from industry, they should represent themselves accordingly to Members of Congress. No member of Congress has time to poke around every group's website and divine whose financial interests are actually being represented during their meeting. Again, NASS may not be doing anything intentionally untoward. However, there is scant transparency in their practices and many potential conflicts of interest, so we are left to scratch our heads and wonder how NASS's parochial drivers affect our nation's efforts to combat Putin and his attacks on our democracy.

Back at the DEF CON report release, the NASS staff's concerns with our speaking lineup were intensified as they became insistent our claims about security were false. The NASS staff then accosted some of our panelists after the event and incredulously claimed that none of these "NSA and military people understood what they are talking about" because the machines and databases are "air gapped" and "unhackable." I remain bewildered as to how NASS staff can claim to know more about security than the men and women who have dedicated their entire career to the security of this nation. I am further mystified as to how they can claim that Chinese-made parts present in nearly every election machine do not create a supply chain threat enabling an attack to be nationally remotely scalable. Moreover, at the event the NASS staff insisted repeatedly that the voter registration databases were secure because they were not connected to the internet. As described earlier, this claim is not solely technically inaccurate, but the Mueller Report explicitly stated that the Russians attacked voter registration databases in the United States via remote internet attacks on the secretary of state websites. So NASS staff were berating a group of national security experts, intelligence officials, and hackers that they did not understand elections and so were wholly inaccurate in suggesting voter registration databases could be attacked via the internet. When, in fact the national security experts, intelligence officials, and hackers were right and remote attacks via the internet are *exactly* how the Russians hacked the voter registration databases.

I should have known then that the NASS staff's belligerent reaction to the DEF CON vulnerabilities report portended more attacks and controversy down the road. We naively believed that vendors and election administrators would want to know what flaws existed in their equipment so they could fix them. What we didn't see coming—that we should have after the NASS staff completely disrupted our announcement—was that NASS, and the largest vendor, Election Systems and Software (ES&S), saw our report as an attack on them.

ES&S was, understandably, determined to protect their enormous share of the election equipment market; as of 2018, 41 states used ES&S products.[7] They didn't seem to understand the cyberthreats or geopolitics at play here as they were still touting election equipment as "air gapped" even after the 2016 election; they myopically viewed our report as an unwarranted attack on them. Had they done a cursory search on Google, they would have found many industries have a symbiotic relationship with hackers. Other industries heavily recruit hackers from DEF CON and conferences like it to work in their security shops. They set up "bug bounty" programs to pay hackers for finding flaws in their system. They bring new devices to DEF CON to be tested before they are rolled out. Unfortunately, this was not the approach ES&S and NASS wanted to take.

Interestingly, we did a reboot of the DEF CON report release at the University of Chicago and added the local head of elections, Noah Praetz, to the agenda to highlight his incredibly impressive security plan for Cook County elections. Noah's reaction at the event was completely opposite that of NASS or the vendors. He was appreciative of the research DEF CON was doing and happy we were proactively highlighting solutions, like his security plan, and not just pointing out problems. Other election administrators at the Chicago event were similarly appreciative of our efforts, and generally wanted to find ways to take what they learned from us and apply what was useful to their systems. We would see this play out over and over again. Local election administrators wanted to work with us and were happy an independent third party was finally doing security research on election equipment. But NASS and the vendors were going to fight us every step of the way.

Determined to change this, we started thinking about the next DEF CON Voting Village—still months away—and how we could improve Year 2. We expanded our plans to invite individual election administrators to participate in DEF CON, and teach the hackers about elections while the hackers researched and disseminated information about flaws in election systems. I raised $25,000 from the Michael and Paula Rantz Foundation to build a database of contact information for about 6,600 election administrators (all we could find online) and paid for a mailing, e-mails, social media, and even 3,500 live calls to their offices. In these communications, we invited them to DEF CON and offered free admission if they brought equipment for us to research.

This invitation was both a genuine olive branch to election administrators and a strategic political maneuver I learned from working with organized labor in democratic primaries.

For example, it's often the case that the national D.C. bigwigs (lobbyists and political leadership) of organized labor unions support a candidate other than yours in a Democratic primary. Once you know your candidate isn't going to get the endorsement of the national union, you go directly to the local chapters or members of the union, and begin communicating directly with them to gain support for your candidate. If you have a good message, invariably some of the local chapters and certainly individual members will side with your candidate instead of the political bosses in D.C. It's a slow process; however, once you have a few champions within the union, they can start working other members directly for you, and before you know it, you may have more union members supporting your candidate than the national leadership has supporting theirs. Judging by the NASS staff outbursts at our Atlantic Council event, we were highly unlikely to get any sort of support from the national association. So we went directly to the membership with our message.

At the same time that we were trying to message directly to election administrators and vendors, we also began thinking about how we could expand beyond voting machines to include the other components of the election infrastructure. As discussed previously, if the machines are at the center of the infrastructure, then the opposite ends are the voter registration databases, and the websites that report the results. One could think of the database that keeps the list of voters as the beginning of the infrastructure. It is where a voter resides before she votes, while the government website that reports the vote totals is the opposite end of the infrastructure. It is where her vote is displayed as part of the vote totals the night of the election.

It was well documented that Russians targeted dozens of voter registration databases in the United States. But it was not as widely known that the Russians had already attacked government websites that announce the election results. As was discussed previously in Ukraine in 2014,[8] Russian hackers attacked the websites that announced election results after the election. Fortunately, the Ukrainians caught the attack and took down their website.[9] Unfortunately, the Russian information warriors had more in store for them. While the Ukrainian website was down, the Russian media began announcing that the Kremlin's preferred candidate, Dmytro Yarosh, won despite the fact that the opposite was true. The Kremlin-backed candidate had lost.

If these two attacks were taken in succession, one can envision a nightmare scenario where Russian hackers attack our voter registration databases and delete enough voters to cause even longer lines than usually occur on Election Day but not enough to invalidate the election—maybe 10 to 20 percent of registered voters. Enough to make the average time it takes to check in a couple minutes longer than usual, because every fifth or tenth person has a debate with the poll worker as to why he is not on the approved list of voters. So on Election Day thousands of people are interviewed on live TV about being disenfranchised because either they were turned away from the polling place or the line was so long they couldn't wait to vote.

Now imagine that after media outlets broadcast the chaos throughout the country on Election Day, Russia hacks the websites of Florida and Ohio. Each state takes their websites down on election night. The country, already uneasy about the reports of chaos throughout the day, are glued to their televisions watching Wolf Blitzer lose it on live TV. Then, as if on cue, Russian Television (RT) announces their preferred candidate has won. Immediately, every fringe media from that candidate's side would run with the news he won. The opposition fringe would try to scream louder that it was all part of a Russian conspiracy. The respective fringe media from both sides and their activists would get worked up into such frenzy they would begin demanding the more mainstream arms of their media outlets cover their version of the events. Pretty soon, MSNBC is covering what only the left fringe was saying a few hours earlier. FOX then toes the line of what the fringe right was saying a few hours earlier. A fiasco of this magnitude would compel the political fringes to become deeply entrenched that their candidate had won and political opponent was attempting a cyber coup. Unwinding the mess and getting to who actually won would take months and make the chaos surrounding Bush versus Gore look like well-ordered democracy.

Again, this is not just my dystopian nightmare: All of these things have, separately, happened. As discussed in previous chapters, multi-hour, obscenely long lines on Election Day are standard in many parts of the country. It is well documented that the Russians hacked into dozens of voter registration databases throughout the country where Russians stole tens of thousands of voters' records.[10] And, we didn't know this at the time of the report's release, the Mueller Report disclosed that the Russians got into the voter registration databases by hacking the websites of the secretary of state offices. So in fact the Russians have hacked government election websites in the United States as well. As described earlier, the Russians hacked the Ukrainian government websites that reported the election results and then erroneously announced their preferred candidate won.

It is also well detailed that Russian bots feed divisive fake news into the bloodstream of the far right and far left. For example, Russian generated fake coverage of the Occupy Wall Street movement on the left as well as gun-rights school shooting issues on the right are consistently picked up and rebroadcast by fringe media on either side and routinely wind up on mainstream media of both sides.

The *only* aspect of this scenario that is new is having all these pieces happen on Election Day in the United States. Every piece of this equation has already happened, and most of it has taken place on Election Day somewhere.

We believed our charge for the next DEF CON, in August 2018, was to try and recreate a complete election system. Since voter registration databases were already attacked in the United States by Russia, we decided to provide free training defending a real list of registered voters for election administration IT staff. To our knowledge, this was the only, hands on, voter registration database defensive training of its kind ever offered.

Recreating the voter registration databases would be easy since it's cheap or sometimes free to acquire the list of voters from a state or county. In fact, it's free to download the most updated list of voters from the Ohio Secretary of State's website. There is some unique software that many of the databases run on but the core of the software is generally Microsoft or one of the other major commercial vendors.[11] So recreating a voter registration database in our cyber range for election administration IT staff to train on was a no-brainer.

However, the websites would be much harder. There was no way in hell even the most sympathetic secretary of state was going to let us hack their website. We would have to build our own. Building websites that looked like a secretary of state website was pretty straightforward. The problem was hacking a website is so easy it would just seem like a "gotcha." And it would be so easy to hack these websites that the real hackers at the conference would not find it interesting. They thought hacking websites was interesting 20 years ago. Remember DEF CON is not a policy, election security, nor even a cybersecurity conference. It is a hacker conference. If you are not doing things interesting to hackers, you will not get invited back.

I mulled over this topic for months. How would one highlight our vulnerability to fake election results, the ultimate fake news? Further, how would one do it in a way that is interesting to hackers?

I stumbled upon the answer while in San Francisco at RSA. RSA is the largest cybersecurity conference (not hacker but security conference) in the world. I made it my side project to ask everyone I spoke to what we should do on the website hacking conundrum. Person after person was stumped. Then, the day before I left, I was sitting on a concrete ledge behind the conference venue with Kevin Collier, then reporter for BuzzFeed, so we could have a conversation away from the mobs of people. I explained my dilemma to him. After thinking about it, he said something like, "Yeah, hacking a website is child's play. The hackers at DEF CON will be totally uninterested in doing that."

I was like, "Right! It's *child's* play! We will have the kid hackers at DEF CON do it!" The first DEF CON was held in 1993. Most of the original founders of the conference, including Jeff Moss, had grown up and had kids. Many of these gray-beard hackers, as well as people like me, want to expose our kids to hacking at an early age. In 2011,[12] a woman named Nico Sell came up with a fantastic idea of setting up what would later be called r00tz Asylum. Like the Voting Village, it's a conference within a conference. In this case it's a conference to introduce children from the ages of five to 18 to the world of hacking. My kids went for their first time at the age of six. I think Moss's daughter got to go before her first birthday. There are multitudes of things kids can learn to hack: locks, drones, and even a Capture the Flag (CTF) game created by Facebook.[13] They can win prizes, learn how to make things with a 3-D printer, and listen to other kid speakers teach them how to be an ethical hacker.

I had felt from the beginning that a central component of the website vulnerability narrative was that it is so easy to hack, adult hackers would find it boring. By extension, that meant Russian hackers would find these websites to be one of the most appealing targets. They could sow maximum chaos with minimum effort. R00tz Asylum provided the perfect vehicle to ensure some hackers at DEF CON (the kids) found the hacking interesting. It also allowed us to raise the alarm for policymakers on Capitol Hill that glaring yet rarely discussed vulnerabilities persisted.

We recruited Brian Markus to ensure the exercise was both educational for the kids and realistic enough to sensitize the public to the threat. Brian is easily one of the best hackers and security people on the planet, bar none. His cybersecurity credentials include being the former Chief Information Security Officer for one of the top rocket manufacturers on the planet, Aerojet Rocketdyne. He develops both security and hacking training for DHS and the Department of Defense. His hacker bona fides include running two of the most popular attractions at DEF CON. The best hackers on the planet come to his DEF CON villages to compete for bragging rights at besting their peers in the challenges he creates. Equally important was the fact that Brian and his team possess an uncanny eye for exquisite design from both an interactive and visual perspective.

We needed these websites to look and feel like the real thing—what voters actually see when they go to their secretary of state's website. And we needed the process of hacking the site to be educational for the kids while at the same time realistic. Then we needed to decide which government election websites made the most sense to rebuild. I quickly settled on the thirteen presidential battleground states. Rebuilding these sites to look indistinguishable from the real thing was fairly easy for Brian's team; however, deciding which type of hack to use was more difficult. Ever committed to art imitating life, Brian had my team research what type of hack was most used against websites in general and election websites, in particular. This exercise quickly landed on an obvious candidate: SQL injection. In plain English, an SQL injection is a basic attack vector that works by entering SQL (Structured Query Language, a standardized computer language) statements into a website's input field, which causes a poorly secured website to execute the SQL statement enabling a hacker to view database information not normally accessible by non-administrators, and that access enables the hacker to modify or destroy data as well as access administrator privileges.[14] Put another way, an SQL injection attack is akin to you asking a stranger, "How are you," and they reply, "Give me your bank account number." A well-secured brain would be confused by that reply, but only a flawed brain would respond with a bank account number.

So Brian created an SQL injection training manual and we recruited some adult coaches like Ricardo Lafosse. Ricardo had recently left government service where he was Cook County's chief information security officer, a role in which he helped design the county's early election infrastructure security programs. Ricardo and others offered to coach the kids on how to hack into the sites using a SQL injection attack. "Operation 'Child's Play'" was chugging along at full tilt.

• 10 •

Cyber Politics

Before I rush into the next DEF CON, I want to pause to consider something Harri and I talked about during our conversation earlier in the book. Harri mentioned being a part of key moments in the development of the internet. He was a part of connecting Japan to Europe, connecting the first 5 million computers to the internet. The first 10 million. The first 20 million.

Since then we have been connecting all the rest of our stuff (homes, phones, cars, refrigerators, medical devices) and about 4 billion people to the internet. It's one thing when we access the internet and by extension, cyber space and cybersecurity from a desktop computer that we get up and walk away from for most of the day. It's another thing altogether when our most personal devices and in an increasingly personal way, we as humans are connected to the internet.

We have moved past connecting just desktops and routers. I contend that in the march to connect all 7 billion humans and the things that most deeply touch our lives like our pacemaker, our baby monitor, and our voting equipment, we by definition are making the internet more inseparable from what it means to be human. The internet is becoming akin to complex language and science: it is unique to humans but at the same time coming to be one of the defining characteristics of humanity.

Making the internet inseparable from humanity, by default, makes the internet more political. As Aristotle said, "humans are political animals."[1] So too, as the internet becomes more inextricable from humanity, it necessarily becomes political. It follows then that as cybersecurity is in part the security of the internet and all the things connected to it, cybersecurity will by necessity become deeply political as the internet continues to envelop humanity.

What we saw in the 2016 election was the politics of cybersecurity, cyber politics, splashing out of the confines of the internet and tech community and into the capital "P," partisan political arena. Three of the most consequential political players in the modern era: Trump, Putin, and Hillary were enmeshed in a struggle for primacy that played out in a cyber narrative. This was surely the most high profile example of cyber politics manifesting itself. It was not the first example, and below I will highlight what I witnessed as subsequent resulting manifestations of the politics of cybersecurity working through this new normal.

Okay, back to DEF CON . . .

Everything was humming along with DEF CON planning when the aforementioned letter hit the mailboxes of election officials across the country. We were inviting them to DEF CON and suggesting they bring equipment for us to inspect. After that letter hit, I began receiving all manner of threatening phone calls from the vendors, their attorneys, and lobbyists.

I was legitimately spooked after these bizarre calls, during which I felt like I was talking to mobster thugs trying to intimidate me to shut down DEF CON Voting Village. Not to be deterred, the next day I reached out to a friend of mine, Ian Bassin, who worked for the White House Counsel Office in the Obama administration.[2] He clearly understood the gravity of their veiled threats and connected me to the Harvard Cyberlaw Clinic. There I met some fantastic attorneys, like Kendra Albert. The Harvard Cyberlaw Clinic agreed to represent us as a pro-bono client. The vendors' threats were growing bolder by the day. The vendors were clearly gearing up for a fight. We also got word from other channels that they were sending staff to DEF CON to identify the serial numbers of the machines at the conference. We were worried they would track down where the machines came from and possibly take legal action against whoever had given us the machines.

My Voting Village cofounders and I were talking about all this during dinner the night before DEF CON with some of the great election administrators who showed up at the conference to speak on our panel and check out the research and other activities. Some of the best and brightest election officials in the country like Neal Kelley from Orange County, California; Noah Praetz from Cook County, Illinois; and Amber McReynolds from Denver, Colorado, were there talking to us about what innovative plans they were putting in place to secure our elections, as well as listen to what the hackers had to say about vulnerabilities they had identified. If the initial shoe-dropping was calls from the vendors about suing DEF CON, the other shoe dropped in the middle of dinner on DEF CON Eve.

In an oddly "coincidental" sequence of events, NASS, in its role as attack dog for the election industry, sent out a letter bashing us and essentially saying our research was illegitimate and unhelpful. They used the many arguments debunked earlier in the book, for example, the environment we hack the machines in not being "realistic" and that they have protections in place to stop someone from getting access to the machines. If that wasn't bad enough, within a few hours of NASS releasing their letter bashing us, ES&S then "coincidentally" released their own letter making nearly the exact same arguments; but, in their letter they also suggested that what Voting Village was doing was unlawful.

We were caught off guard, stunned, and I, in particular, was livid. So far the NASS pseudo-lobbyists had made a huge scene at our report rollout event a few months earlier. Then the vendors threatened me with their lawyers and lobbyists a few weeks prior. Now there was what felt like a surprise attack on us on DEF CON Eve, attacking our credibility and, in ES&S's case, threatening us with lawsuits. Enough was enough.

This series of events and public threats was the most high profile mishandling of their cyber politics from NASS and the vendors to date. In a pre-cyber politics world, they could just bully everyone into submission. But in the new normal of cyber politics they had stepped into a hornets nest and the press, social media, and notably other key policymakers, like those in Congress and on the campaign trail, took note of the letters for what they were: cheap shots that were factually inaccurate, needlessly defensive, and tone deaf to boot.

I had worked on my speech for the opening of DEF CON for two weeks. It was all "kumbaya" about how great it was that the research community and election administrators were coming together, and how great it was to have so many election administrators here. But because of the letters the night before, I threw it out and started over the next morning. I walked into the room and welcomed everyone. I thanked the many election administrators in the room who had come to DEF CON to speak to the attendees and learn from the research. I praised the election administrators in attendance for their deep commitment to security and protecting votes from foreign attack. Despite this being an antijingoistic crowd, I even said the election administrators were "great Americans" who were deeply patriotic and care about their country. Then I held up copies of the two letters from NASS and ES&S. I pointed out that no one had the courage to sign either letter. I said that I was not speaking for DEF CON or the Voting Village, just myself. I stated, "Just one measly human out of 7 billion humans. Just standing here in jeans and a T-shirt to say *Fuck. You. You fucking Luddites*." The room erupted. Everyone was cheering, standing on their feet, and hooting and hollering.

To be clear, no one was cheering for me. Most of the people in the room had no idea who I was. Rather, they were releasing a collective rage against the very groups who, for two years, had been rejecting the *free* help we were offering to secure their deeply insecure systems. They were reacting to how little had been done to secure our elections since 2016. They were reacting to the unwillingness of the election industry writ large to take the threat seriously and admit how vulnerable it was. Like the founders of DEF CON, the participants felt ambushed by the letters on DEF CON Eve, and they were reacting to that, too. Maybe more than any other group of hackers at DEF CON, those involved in the Voting Village feel like they are providing a public service. They felt kicked in the teeth by a group of nameless, faceless D.C. lobbyists who didn't understand technology, national security, or hackers.

Then, out of nowhere, we received an unlikely supporter: Trump's former White House Cyber Czar, Rob Joyce, was speaking at DEF CON and took the ES&S and NASS letters head-on. He said in public remarks that the officials who were criticizing us needed to realize that "believe me, there are people who're going to attempt to find flaws in those machines, whether we do it here publicly or not."[3] Obviously as a top NSA official, he wasn't talking about corrupt city council candidates, he was talking about foreign actors trying to hack our elections. This was not only unexpected, but also a huge political buffer to our efforts. To have someone from the Trump administration backing us took away a key line of attack. Now it would be hard for anyone to argue that what we were doing was a veiled attack on the president. Moreover, this is the first time a senior NSA official publicly denounced the dismissive stance of the election lobbyists. A rebuttal from someone like Joyce was clutch in fighting their claims. We began citing it in every interview and pushed it hard on social media. Cyber politics was beginning to play out and we were milking it for everything we could.

Just a few minutes later, the first kid hacked into the simulated Florida Secretary of State's website. Between introducing speakers for the Voting Village, I ran down to r00tz Asylum to see what was going on, and it was a feeding frenzy for the media. I knew it would be popular with the media, but I didn't realize to what extent. There were about a dozen kids sitting there hacking into the first website. After looking at it, I realized part of the appeal was the juxtaposed visuals. From one perspective, it looked like some kids doing routine school work with a few adults coaching them along. They were sitting with instruction manuals and old laptops. The first camera shot is an idyllic scene of cute kids and their parents looking through some papers and innocently typing on an old laptop. Essentially the same scene that plays out at homework time in millions of homes across the world. The next camera shot is of a huge screen with a website that looks exactly like the real Florida Secretary of State's website, and the vote totals and even candidate names are being changed. The shot moves from this familiar, idyllic picture of kids learning with their parents to a nightmare hellscape where faith in our democracy is undermined with the keystroke of a child. These types of scenes are exactly what the press lives for and they pounced. Within minutes of the kids sitting

learn, SQL injection attacks are the *exact same attacks* the Russians used on the secretary of state websites in 2016. We now know this thanks to the Mueller Report.[7] In the report he states Russians used SQL injection attacks to breach the websites and hack voter registration databases.[8] Hacking the website and then hacking the voter registration database is far more complex than just hacking the website to deface it, as the kids were doing. Let me just state that again: Many secretaries of state and their apologists adamantly denounced us on social media, in the press, and to senior policymakers. They vehemently insisted an SQL injection could not successfully hack their website. However, an SQL injection is the *exact* attack the Russians used to hack their websites. In retrospect, these public attacks on DEF CON may have betrayed a substantive problem with efforts to secure election systems since 2016. Recall from earlier in the book that the secretaries of state all had been granted security clearances by now. The federal government had gone to great trouble and expense to grant the clearances. So while DEF CON attendees had to wait until the Mueller Report came out to know the Russians used an SQL Injection to hack their websites and voter registration databases, the secretaries did not. In fact, DHS had gotten the secretaries clearances *specifically* so they could learn what attacks were used against them and remediate their systems. It's possible the secretaries knew SQL Injections were used against their websites and just lied and attacked us anyway. However I highly doubt that's the case. The cost-benefit analysis of getting caught in a lie would likely preclude them from that approach. It's also possible they found out SQL Injections were used against them in 2016 and subsequently bought security tools to defend against SQL Injection attacks. I hope this is what in fact happened. But, if they had this level of understanding, they likely would not be so adamant they can stop an SQL Injection, as the defensive tools available are far from perfect. And as it is human nature to toot your own horn, at least one of them likely would have proclaimed that as part of their robust security response to 2016 they had new tools and procedures in place to secure their systems. Further, if they knew about the SQL Injection attack and then took steps to remediate it, it would seem at least one of the vocal secretaries would have said that while they are defending against these attacks, they hope local election officials will heed our warnings, as these locals' websites were vulnerable to such attacks. But none of that happened; unfortunately their public attacks on DEF CON likely uncovered that they had not bothered to ask DHS what attack was used to hack their websites and thus databases. This is no small oversight. The one thing we *know* the Russians did in 2016 was attack our voter registration databases. As stated earlier in the book, these databases are central to administering the election. Hacking them can cause a number of catastrophic problems in the election, like deleting voters from the list so they can't vote or causing long lines. Either of these attacks would create visible chaos on Election Day that could undermine confidence in the election, disenfranchise voters and alter vote totals and therefore winners and losers. So we would expect one of the first questions the secretaries would ask upon receiving security clearances was how the databases were hacked. At least for those who were attacking DEF CON, it seemed they did not know this attack was used against them. This was arguably the top security risk to elections after 2016. It seemed two years later this low hanging fruit to secure against a known threat had yet to be mitigated against. If this is the case, it does not bode well for what other security has been implemented since 2016.

After observing the kids for a minute, I literally ran back to the room with the speaking track because I was late to moderate what was certain to be a contentious panel. This is when naked politics began to permeate the rest of the conference. Not partisan politics. Rather, individuals using different aspects of cybersecurity to further their own political agenda, for instance, cyber politics. The panel included several state and local election administrators as well as Jeanette Manfra, who was Trump's deputy assistant secretary from DHS charged with overseeing election security; and California's Secretary of State Alex Padilla, who had been an ardent critic of the Trump administration on election security. The panel was relatively tame until the last question.

A hacker sitting toward the front of the room raised his hand and asked if the Trump administration should put Putin on notice that if Russia hacked our election infrastructure, the United States would consider it an act of war. He was clear to make the distinction that things like fake news didn't rise to the level of hacking election infrastructure, but that we should draw a line somewhere and let Putin know we are deadly serious about protecting our democracy.

Manfra was caught off guard by this politically charged question and understandably fumbled the beginning of her answer while she considered the implications of her response. She eventually said, "No, I don't think it's an act of war."[9] I said thank you and moved to end the panel. Secretary Padilla almost grabbed the microphone out of my hand and said,

> One of the things that you, that all of your colleagues will say . . . it takes, what we face in 2016 and continue to face now requires a whole of government response. Right, [to Manfra] that's not classified, you've been saying it publicly. Requires a whole of government response. Last I checked, the person who sits in the Oval Office is a part of our government, and, as great as we [indicating Manfra] are working together, we still need the right words to come out of the mouth of the sitting president of the United States of America, and it has not.[10]

People were stunned by the exchange. I was admittedly sympathetic to Padilla's view and made no attempt to stop him from making his comments attacking the administration or cut him off when his answer went long. It was an important policy debate to have between two people at the center of the issue and I wanted to let it play out. Cyber politics was flaring up over a critically important topic between two partisans with opposing political interests to protect. Oh, and it was great political theater too.

After this heated exchange I ended the panel and walked out of the room to take a few phone calls. I immediately began speaking to several candidates challenging incumbent secretaries of state (DEF CON was happening only three months before the 2018 Midterm Elections). Several of them were asking for our help in attacking the incumbents for their support of NASS and/or their state's use of ES&S's equipment. This was new territory for DEF CON and as blatant an example of cyber politics as anything that happened at the conference. Having candidates run for office on a "pro-hacker" agenda was unprecedented. But the media was covering this topic extensively and the hacker community had exposed undeniable flaws in our electoral system. So when the election lobbyists took a position that was obviously self-serving and counter to the security of our democracy, they left huge openings to be exploited for political gain on both sides of the aisle.

Exploited they were. While no one seemed to come to the defense of the election lobbyists we began seeing an outpouring of support from Members of Congress, individual election administrators who attended DEF CON like California Secretary of State Padilla, researchers, national security leaders like Trump Cyber Czar Rob Joyce, and many more than I can name here. Attempting their best Wile E. Coyote impersonation, ES&S sought to step on the rake one more time for a final whack from the political debacle they created for themselves by levying another attack on the last day of DEF CON. The backlash to this attack hit a crescendo when the *Wall Street Journal* posted a story that was largely critical of ES&S,[11] coupled with Rob Joyce,[12] the Electronic Frontier Foundation,[13] and a host of others took to Twitter to denounce ES&S's actions and remind us that "Ignorance of insecurity does not bring you security."[14] I'm challenged to think of another issue that has the NSA, EFF, ACLU, and *Wall Street Journal* on the same side. NASS on the other hand seemed to see the forest for the trees and stopped attacking us publicly. After DEF CON they even took the bold step of e-mailing our generic info@ e-mail address in what was a sort of backhanded olive branch. Critical of us, but outreach to reboot our relationship nonetheless.

So, what did we find that made ES&S irate enough to continue attacking us despite criticism from election officials, Trump administration officials, as well as Democrats and Republicans in Congress? As had been the case the previous year, the number and severity of vulnerabilities of voting equipment still used in the United States discovered in 2018 was staggering. Among the dozens of vulnerabilities found in the voting equipment tested at DEF CON, all of which (aside from the WINVote) are still used in the United States today, the Voting Village found the following:

- A voting tabulator that is currently used in twenty-three states is vulnerable to being remotely hacked via a network attack. Because the device in question is a high-speed unit designed to process a high volume of ballots for an entire county, hacking just one of these machines could enable an attacker to flip the Electoral College and determine the outcome of a presidential election.
- A second critical vulnerability in the same machine was disclosed to the vendor a decade ago. Yet that machine, which was used into 2016, still contains the flaw.
- Another machine used in eighteen states was able to be hacked in only two minutes, while it takes the average voter six minutes to vote. This indicates one could realistically hack a voting machine in the polling place on Election Day within the time it takes to vote.
- Hackers had the ability to wirelessly reprogram, via mobile phone, a type of electronic card used by millions of Americans to activate the voting terminal to cast their ballots. This vulnerability could be exploited to take over the voting machine on which they vote and cast as many votes as the voter wants.

That's right, a machine used to count millions of votes in twenty-three states could be hacked remotely. Breaching just one of those machines could flip the Electoral College, and thus the presidency, and *ES&S has known about it for ten years and still hasn't fixed it.* No wonder they were skittish and threatening us in public. That seems to strike a mortal blow to their claim of taking security seriously.

As I was taking off on a flight from Chicago to Des Moines about a week later, I was on the phone with a reporter who was asking some questions for a follow-up story. He stopped mid-sentence and said, "Holy shit. Did you see the Senate letter that just dropped?" Right at that moment, the voice came over the loudspeaker saying we all had to get off our phones. I said, "What?!?! Can you send it to me? I don't see it." He said, "Sure." And we hung up. I frantically searched any letters coming out from the Senate that mentioned election security until I lost cell service from the altitude. My stomach continued to churn, and for the next hour I sat there with my mind spinning trying to think of what horrible thing the lobbyists and lawyers could have conjured up to undermine us. Maybe a letter denouncing DEF CON? Saying we needed to cease and desist from hacking machines? Calling us in for testimony or whatever other D.C. parlor trick the lobbyists could come up with to stop us? It was never lost on me that we were still a bunch of volunteers and had real jobs to get back to. These guys had a lot more money, and this was all they did. This was their job. All day. Every day. In that equation, the volunteers usually lose. Someday these guys would figure out how to play cyber politics too. Maybe this was the day.

Once I landed, the reporter's text with a link to the letter came through, and the people sitting around me must have thought I just had a stroke. In a move that floored all of us, Senator Harris (D-CA) organized a bipartisan group of leaders on the Senate Intelligence Committee: Senators Harris, Warner (D-VA), Lankford (R-OK), and Collins (R-CT) took the unprecedented step of sending a letter to ES&S demanding they stop harassing DEF CON and work with the hacker and research community to secure our elections:

> We are disheartened that ES&S chose to dismiss these demonstrations [by DEF CON Voting Village] as unrealistic and that your company is not supportive of independent testing. We believe that independent testing is one of the most effective ways to understand and address potential cybersecurity risks.

No one in Congress had ever given this much support to hackers. Much less stand up to election lobbyists and lawyers and tell them to lay off the hackers. What in God's name was going on? Cyber politics is what was going on.

Senator Harris is from Oakland, neighboring San Francisco and Silicon Valley, two of the most tech-obsessed places on the planet. Oakland is no slouch when it comes to tech start-ups either. She also represents a state that likes to think of itself as the global center of innovation, whether it be culture (Los Angeles) or tech (Silicon Valley). Harris is a savvy politician and an astute policymaker. She realized that her constituents cared deeply about the integrity of our democracy, and they were also technologically advanced enough to see through the erroneous claims of the lawyers and lobbyists.

The move astounded many in the election community, not least of which was us. In conversation with the Senator's office, they were very open about the fact that they had wanted to "do something" about this issue before. Politics aside, the Senator and her team care deeply about this issue. It wasn't until every major media outlet in the country was covering ES&S's massive political cyber sin of practicing ignorance of insecurity that they felt there was room to step into the conversation.

ES&S's response to the bipartisan Senate letter was so politically blundering it was comically bad. It dripped with D.C. lobbyist double-speak and contempt. In an astounding rebuke to the senators at the end of the letter, despite being asked to work with DEF CON and stop attacking us, something they ostensibly agreed to at the beginning of the correspondence, they called on the Senate Intelligence Committee to have federal investigators *investigate DEF CON* because we allowed foreigners to hack the machines. ES&S was appropriately mocked on Twitter[15] and came off as part of the problem as opposed to the solution. Neither we nor any of the senators felt the need to respond to their letter.[16] At the same time, it certainly did not inspire hope that they had gotten the message and were on a path to improve.

When the Lobbyists and Lawyers Win, We all Lose

While ES&S was executing this masterful faceplant in Congress, NASS, was lobbying for something in Congress far more sophisticated and insidious. They were quietly killing the leading bill in the Senate that had anything approaching real security standards for voting equipment and systems. And again, since they don't have to register as lobbyists, no watchdog groups or journalists could shine a light on this activity. Senators Amy Klobuchar (D-MN) and James Lankford[17](R-OK) did their best to fight off the lobbying. Senator Klobuchar even said publicly in a moment of exasperation,

> As you can imagine, we've spent days and days with the local secretaries of state, and some of them just aren't going to like that we're requiring these backup paper ballots if they get federal money. And these audits, I just don't know what we're going to do, and the members [of Congress] are going to have to rise above that in my mind.[18]

Her pleas were to no avail. The voters don't have lobbyists in D.C. to counterbalance the vendors' and NASS's lobbying. So, once again, the voters lost and the lobbyists won.

Since 2016 the record of the vendors and NASS lobbyists was:

1. Actively work to publicly discredit the only public, third-party assessment of voting equipment in the world (DEF CON) and threaten us with lawsuits.
2. Remove common sense spending restrictions on the $383 million in Congressional funding for local governments to secure their elections.
3. Kill the bipartisan bill in the Senate that ensured practices all experts agree would safeguard elections, like basic cyber hygiene, paper ballots, and audits.

When Cyber Politics Play Out in our Favor

However, not to be deterred as all this was transpiring, Congresswoman Jackie Speier took the bold step of asking us to release the 2018 DEF CON report in the U.S. Congress. This was the new world of cyber politics at play yet again. Representative Speier's district covers parts of San Francisco and Silicon Valley.[19] She genuinely cares about this issue and has been talking about it for years; but elevating the hacker perspective on election security is also good politics for her with the constituents in her district. The tech community, similar to the national security establishment, understands how grave a risk "head in the sand security" is for our democracy. Representative Speier may hail from the most high-tech district on the planet. Her constituents understand in a way many of us never will how deeply insecure IT systems and, for that matter, ten-year-old voting machines that have never been patched are. They not only welcome her leadership on this issue, they expect it.

And we had gone from late 2016, when the public and policymakers were essentially taking the word of the election lobbyists that our elections were not only "air-gapped" and "unhackable," to a consensus that the election equipment was so insecure that paper ballots and audits were our only hope to stave off a successful attack. We had also helped position election security as a national security issue, not an election administration issue. What makes us even more hopeful for the future is that the political odds are more in our favor now. Senate Republicans have, by and large, always been good on this issue. Senator Lankford (R-OK) has shown particularly strong leadership on election security. The stumbling block since 2016 was House Republicans who the election lobbyists could easily gin up to stop progress. With a Democratic House and a Republican Senate, both of whom want to fix this issue, we stand a chance against the election lobbyists. We see this positive bipartisan political dynamic playing out already with the House Homeland Security Committee proposing a bill on election reform. The bill has a strong section on election security that my cofounders at DEF CON and I gave input on during drafting, and we were asked to testify to the Committee about the merits of the bill. Obviously this was the first time anyone from DEF CON has testified to the House Homeland Security Committee on the "hacker perspective" of an election security bill. Early reports show that House and Senate Republicans like the security section of the bill and may be willing to pass that section in the future. On top of that, the National Academies of Sciences, Engineering, and Medicine came out with a truly comprehensive assessment of what voting security should look like, which includes all the things already mentioned in this book and a lot more.[20] So since 2016 hackers and researchers have:

1. Founded the only public, third-party assessment of voting equipment in the world.
2. By highlighting the vulnerabilities in voting equipment have helped secure $383 million in funding for election security upgrades.

3. Provided impetus for a bipartisan letter from Senators castigating the top election equipment vendor for not accepting security advice from the research community. This, in part has led many of the vendors to accept third-party assessments from other researchers.
4. Advised the drafting of a sweeping bill in the House to secure our elections.
5. Testified in Congress as to the merits of common sense security and worked with both chambers of Congress and both parties to begin passing a bill mandating security.

These successes suggest a positive trend. This level of activism in politics and legislation by the hacker community was unheard of just a few years ago. Further, hacker activism being taken seriously by policymakers and leading to demonstrable change was unthinkable until recently. As the internet becomes an inextricable part of what it means to be human, the people who can find and fix the security flaws in the internet and everything that touches it will necessarily play a larger role in our political discourse. So despite the best efforts of the lobbyists and lawyers, the politics of cybersecurity may prevail in favor of the voters in the long run.

Epilogue

Hopefully this book has made the reader aware of the glaring challenges we face as a nation in securing our elections from cyberattacks by Russia and other nation-states or determined adversaries. A few tenacious hackers and political operatives, as well as some deeply committed election officials, worked tirelessly to repudiate the false claims by vendors and some election officials after the 2016 attacks that our election infrastructure was "unhackable" and "air-gapped." While the hurdles to fix these problems are real and will require an enormous amount of determination and national focus to remedy, I am certain that the United States will eventually rise to the challenge and stave off further attacks to our election infrastructure. We will likely be stuck with fake news and the like forever; but ultimately, we will prevent foreign nations from hacking our voting machines, registration databases, and so on, if for no other reason than the national security establishment is bound to get fed up with these attacks and enact punishments severe enough that they deter Russia from doing it again, and quickly punish anyone who attempts an attack.

There is possibly a far brighter silver lining to this episode in U.S.–Russia relations. In a dramatic stroke of irony, Putin may have initiated a series of events that could lead to the strengthening of election processes in developing democracies throughout the world, which could ultimately loosen the grip of the very despots with whom he hopes to ally.

Unlike in the United States, where there is no evidence of systemic voter fraud in modern elections, in many developing democracies there is legitimate concern of widespread tampering with the ballot box. In fact, one serious concern from 2016 is that copycat despots, in places like Turkey, Egypt, Honduras, El Salvador, or elsewhere, will use these same practices not against their enemies abroad, but to steal elections domestically. In the United States, seventeen intelligence agencies are constantly on guard trying to identify if the Russians or anyone else are interfering with our elections. In most developing nations, no one other than groups like the Carter Center monitors elections. The Carter Center has observed 107 elections in thirty-nine countries, such as Bolivia, Egypt, Peru, Libya, Lebanon, and Nicaragua, since 1989. Needless to say, these international nongovernmental organizations (NGOs) have scant cyber capability and little to no chance of uncovering a cyberattack on an election.

This highlights a broader downside to the Fourth Industrial Revolution, discussed earlier in the book, as we connect all the things we make and the last 3 billion humans to the internet. This convergence of connectivity poses the greatest risk to human rights in the next century. Period.

And the particular human rights at risk are those of the last 3 billion people themselves.

Silicon Valley would have you believe that total coverage of the internet is necessarily good, and that once everyone in Somalia has 5G internet access they will all watch Khan Academy videos, become autodidact learners, and Utopia will ensue.

There is no doubt that access to the internet, more data, and the ability to gather data and quantify their lives can provide a lot of opportunity for those who are yet to come online.

But it is not necessarily the case that the internet is all good all the time. Technology is neither good nor bad. It is neutral, and the people who use it decide whether they are going to use it for good or bad.

For the past 30 years the internet has spread, by and large, through countries with robust civil society institutions, rule of law, and checks and balances on government overreach. Unfortunately, the last 3 billion are the last 3 billion for a reason. They live in countries who have both figuratively and, quite literally, not enabled these citizens to "plug in" to the modern world.

We should continue to celebrate inspiring stories of how the vast network of NGOs, politicians, and visionary entrepreneurs in the developing world use the internet for good. But we should similarly be very clear-eyed that the despots, criminals, and warlords oppressing the last 3 billion will use the internet for evil. And they will use it for a far greater degree of evil against the last 3 billion. Those are specifically the people who have the least power and access to redress, while their abusers are people with significantly more power and ability to undermine their human rights with impunity. It is likely that a key focus of these authoritarian rulers and thugs will be using the internet to impede their citizens' right to vote.

That being said, some innovative international development programs offer a ray of hope. We could not only curtail impending human rights abuses but strengthen developing democracy in the process. A template for this may be the World Bank Identification for Development (ID4D) program. The program fights corruption in the developing world via reformed identification systems, often making use of biometric identity cards or profiles. After years of seeing billions of dollars meant for the last 3 billion people bilked by local thugs and corrupt politicians, the World Bank launched this initiative. It finances the implementation of legal proof of identity for all citizens in a country. So they receive the benefits and rights they deserve, while verifying that these programs are helping those intended. There are a variety of systems in use, with quickly developing international standards to allow for interoperability between nations. The scheme enables each citizen to register her biometrics usually including fingerprints, retina, face, and often voice, with the government. The combination of this information is virtually impossible to fake with present-day technology. Once secured with highly sophisticated encryption, the record is tied to the benefits the person is intended to receive. These benefits could include health care services, education tuition, financial support, food stuffs, and so forth. With use of the ID card—a smart card, a 2-D card with barcode, a SIM card connected with a mobile device, or even a file stored in a central system or cloud—delivery of these services can be easily tracked by senior officials to ensure the intended recipients receive the correct level of support.[1] Furthermore, they can monitor whether funds were grafted by local corruption. Of course, there are elaborate schemes a local thug could use to circumvent this technology and process. To be sure, there are reports of corrupt officials finding ways to thwart the process; however, the World Bank has made it categorically harder for the average local warlord to steal benefits from the last 3 billion. The program's popularity among voters has motivated politicians in other democracies to implement a biometric ID program of their own volition.[2]

Add to the World Bank anticorruption program template a technological development that was stalled for years but, because of Putin, has newfound funding and life breathed into it. Open-source voting technology has been the holy grail of election security experts for decades. One of the biggest gripes of the security community is that voting equipment is only studied publicly by third parties for two and a half days a year at DEF CON. All other security assessments are done behind closed doors, if at all. Neither the public nor policymakers, nor national security cyber experts, have a systematic process to assess the vulnerability of the election equipment. This situation was fine twenty years ago, and is still fine for the security of toasters and night lights; but it is no longer good enough for voting equipment.

Open source turns the old model on its head. Open-source software is exactly what it says it is. The source code for the application is open to anyone to review, audit, and suggest changes to the owner of the code.[3] A great open-source application success story is that of the texting app Signal. Signal is a favorite of the cybersecurity and hacker community specifically because it is open source. The logic goes that no company has enough time or resources to constantly identify vulnerabilities and successful breaches; but if one has an active user base that cares about the security of the app, white hat hackers, hobbyists, and security researchers will spend some portion of their free time reviewing and proposing changes to the code. It turns out that a few hours here and there from enough enthusiasts eventually reach a critical mass whereby their collective efforts are far greater than anything the company could afford to do on its own.[4] This is the same community-improvement principle Wikipedia, eBay, and Craigslist use to police the content, if not the code, on their sites. And aside from the Birkenstock-wearing folks at Craigslist et al., even the notoriously buttoned-up top brass at the Pentagon have started a program called simply "Hack the Pentagon," where they encourage hackers to find vulnerabilities the military can fix and provide rewards to those who find previously unknown vulnerabilities.

This process works. Corporate, academic, and government security and technology leaders endorse the value of open-source practices.[5] How would one put the World Bank anti-corruption program and open-source voting technology together to make progress in fighting election rigging in emerging democracies? First, the World Bank would convene a select group of election leaders from throughout the world and organize a thorough review of election vulnerabilities, once satisfied, it would vote to "endorse" one or more open-source election software that could be used on any tablet or smartphone. The code would then reside in a World Bank owned cloud; but anyone could access, audit, and provide suggestions to improve the code at any time. The project could begin with the countries that have already implemented the biometric ID program. The World Bank would offer to pay for the technological administration of the first three to five elections (a financial burden for most emerging democracies) if they commit to using the open-source software. To further ensure the security of the vote, the World Bank could buy inexpensive tablets that are stripped down and just have the open-source software loaded on them, or have them configured to automatically synch with the World Bank cloud and download the software once they are powered up.

Buying 100,000 tablets in bulk for Nigeria or 20,000 for Honduras would be a rounding error in the World Bank budget. And other NGOs committed to emerging democracy election integrity, like the Carter Center, could chip in as well. Providing the devices for free or with heavy subsidies would be a massive marketing opportunity for Google, Microsoft, Amazon, and their global competitors, driving competition to deliver them with minimal costs. Finally, people would vote on paper ballots that would be scanned and tallied with the open-source software on the devices. Audits could be mandated. The paper ballots would provide a paper trail if audits showed there were any problems with the technology. The World Bank could tie this funding to other forms of funding, for instance, low-interest loans and grants.[6]

This program would have the dual benefit of making the elections more secure from tampering and increasing overall faith in the elections among the voters in emerging democracies because of the internationally approved technology and processes, as well as multiple redundancies, like the paper ballots. As in the United States, many in the developing world are suspicious of multilateral organizations imposing the will of foreign interests on their population. But suspicion of their own government is pervasive as well. So the combined oversight of a third-party multilateral organization and transparency of the underlying code that *any* voter in their community can review is ideal. This program provides such a multifaceted oversight. The World Bank would fund an initial election infrastructure replete with capability to monitor and thwart vote tampering. This would give voters in these countries another level of assurance that their votes are counted as cast and the outcome is accurate. In addition, to voters in their own community, anyone else on the planet could audit the election software code, and reputable groups, like the Carter Center, could further inspect and validate the integrity of the code.

As is the case with so many other advances in infrastructure, it is specifically the emerging democracies, where much of the last 3 billion live, that are in the best position to benefit from such a World Bank program. Outdated laws, entrenched vendors, and legacy systems make such a program impossible in developed democracies like the United States, the EU, or parts of Asia. While much of the biometric ID technology was developed in the United States,[7] it will be decades, if ever, before a national biometric ID program can be adopted here. Challenges relating to vendors who make state IDs, the myriad privacy laws, and revenue streams states depend on from state IDs[8] would doom any such effort here. But, for better or worse, countries like Djibouti, Nepal,[9] and Indonesia don't have these laws, vendors, or constraints in place. They were able to implement robust biometric ID programs with the World Bank's help. If there is political will and a robust capital outlay, emerging democracies can complete huge advances in technology and transparency with respect to elections that the United States, the EU, or other developed countries would struggle to even obtain permission to explore, much less carry out.

In fact, the best example of where these types of advances have already happened is Estonia. After the Cold War, they tried to rethink their government with a blank-slate mentality. Since they were not beholden to any Soviet infrastructure or vendors, they built a government that ran on the internet. This would have been virtually impossible without the ability to start from scratch and build the government infrastructure out of whole cloth instead of atop legacy systems, vendors, and regulations. It is estimated that this so called "e-government" saves the average Estonian a full week of their life from waiting in line at a government office to file forms or other mundane tasks.

Emerging democracies throughout the world have a similar opportunity. The World Bank has years of experience implementing the biometric ID program among billions of people, and recent advancements in open source voting technology have made deployment of open source voting software a near term possibility. We are closer than ever to achieving all three of the following major advances in protecting the right to vote of the last 3 billion people to come online:

- wresting the power to rig their own elections from despots;
- securing elections from a myriad of other external nefarious actors; and
- dramatically increasing faith the last 3 billion have in the integrity of their democracy.

Achieving any of these would be significant progress for the last 3 billion as they and all their stuff comes online. Accomplishing all three could be the greatest advancement to protect the human rights of the last 3 billion in the digital age.

And we would have Putin to thank for starting the chain reaction that made it all happen.

Notes

INTRODUCTION

1. "Voting Systems and Use: 1980–2012," *ProCon.org*, https://votingmachines.procon.org/view.resource.php?resourceID=000274 (accessed March 26, 2019).

2. David E. Sanger, "Putin Ordered 'Influence Campaign' Aimed at U.S. Election, Report Says," *New York Times*, January 6, 2017, https://www.nytimes.com/2017/01/06/us/politics/russia-hack-report.html (accessed March 26, 2019).

3. Chris Good, "How Hackable Are American Voting Machines? It Depends on Who You Ask," *ABC News*, October 15, 2018, https://abcnews.go.com/Politics/hackable-american-voting-machines-depends/story?id=58511054 (accessed March 26, 2019).

4. "Elections Performance Index," MIT Election Data and Science Lab, https://elections.mit.edu/.

5. "Secretary Napolitano Opens New National Cybersecurity and Communications Integration Center," *Department of Homeland Security*, October 30, 2009, https://www.dhs.gov/news/2009/10/30/new-national-cybersecurity-center-opened (accessed March 26, 2019); Spencer S. Hsu, "Obama Combines Security Councils, Adds Offices for Computer and Pandemic Threats," *Washington Post*, May 27, 2009, http://www.washingtonpost.com/wp-dyn/content/article/2009/05/26/AR2009052603148.html (accessed March 26, 2019).

6. The 20 CIS Controls & Resources," Center for Internet Security, https://www.cisecurity.org/controls/cis-controls-list/.

7. Jeff Kauflin, "The Fastest-Growing Job with a Huge Skills Gap: Cybersecurity," *Forbes*, March 16, 2017, https://www.forbes.com/sites/jeffkauflin/2017/03/16/the-fast-growing-job-with-a-huge-skills-gap-cyber-security/#5797043d5163 (accessed March 26, 2019).

8. Tal Parmenter, "IT Students Selected as Interns to State of Illinois Internet Privacy Task Force," *Illinois State University*, December 1, 2001, https://news.illinoisstate.edu/2011/12/it-students-selected-as-interns-to-state-of-illinois-internet-privacy-task-force/ (accessed March 26, 2019).

9. "Aristotle Insists That Man Is Either a Political Animal (the Natural State) or an Outcast like a 'Bird Which Flies Alone' (4thC BC)," *Portable Library of Liberty*, March 17, 2008, http://files.libertyfund.org/pll/quotes/164.html (accessed March 26, 2019)."Jeffersonian Ideology," *U.S. History: Pre-Columbian to the New Millennium*, http://www.ushistory.org/us/20b.asp (accessed March 26, 2019). "Montesquieu, Natural Law, and Natural Rights," *Nlnrac.org*, http://www.nlnrac.org/earlymodern/montesquieu/documents/spirit-of-laws (accessed March 26, 2019).

CHAPTER 1

1. Wolf Blitzer, "Romney: Russia Is Our Number One Geopolitical Foe," *CNN*, March 21, 2012, http://cnnpressroom.blogs.cnn.com/2012/03/26/romney-russia-is-our-number-one-geopolitical-foe/ (accessed March 27, 2019).

2. David M. Drucker, "Romney Was Right about Russia," *CNN*, July 31, 2017, https://www.cnn.com/2017/07/31/opinions/obama-romney-russia-opinion-drucker/index.html (accessed March 27, 2019).

3. "Bush Trusts Putin," *C-Span*, June 16, 2001, https://www.c-span.org/video/?c4485932/bush-trusts-putin (accessed March 27, 2019).

4. Andrea Kendall-Taylor and David Shullman, "How Russia and China Undermine Democracy," *Foreign Affairs*, October 2, 2018, https://www.foreignaffairs.com/articles/china/2018-10-02/how-russia-and-china-undermine-democracy (accessed March 27, 2019).

5. "The Intelligence Community Report on Russian Activities in the 2016 Election," *Washington Post*, http://apps.washingtonpost.com/g/page/politics/the-intelligence-community-report-on-russian-activities-in-the-2016-election/2153/ (accessed March 27, 2019).

6. Presentation by Douglas Lute, October 22, 2018.

7. Allan Smith, "Stacy Kemp Refuses to Call Brian Kemp's Victory 'Legitimate' but Says It Is 'Legal,'" *NBC News*, November 18, 2018, https://www.nbcnews.com/politics/elections/stacey-abrams-refuses-call-brian-kemp-s-victory-legitimate-says-n937721 (accessed March 27, 2019). Mark Niesse, "Without disclosing evidence, Kemp accuses Georgia Democrats of hacking," The Atlanta Journal Constitution, November 4, 2018, https://www.ajc.com/news/state--regional-govt--politics/kemp-office-investigates-georgia-democrats-after-alleged-hacking-attempt/VyyeNgNH4NN6xaXUOfl8HL/ (accessed March 27, 2019).

8. Interview with Douglas Lute, October 22, 2018.

9. "Russian Election Hacking Wasn't as Bad in 2018. That's No Excuse to Sit Back and Relax," *Washington Post*, November 24, 2018, https://www.washingtonpost.com/opinions/russian-election-hacking-wasnt-as-bad-in-2018-thats-no-excuse-to-sit-back-and-relax/2018/11/23/7edcac9a-e2d6-11e8-8f5f-a55347f48762_story.html?utm_term=.4e4514897368 (accessed March 27, 2019).

10. Lucan Ahmad Way and Adam Casey, "Russia has been meddling in foreign elections for decades. Has it made a difference?," *Washington Post*, January 8, 2018, https://www.washingtonpost.com/news/monkey-cage/wp/2018/01/05/russia-has-been-meddling-in-foreign-elections-for-decades-has-it-made-a-difference/?utm_term=.1497522ef970.

11. Nick Espinosa, "Cutting through the 'Fake News': Proof That Russia Is Hacking Everyone," *Forbes*, August 15, 2018, https://www.forbes.com/sites/forbestechcouncil/2018/08/15/cutting-through-the-fake-news-proof-that-russia-is-hacking-everyone/#3180248378f0 (accessed March 27, 2019).

12. Tyler Whetstone, "Knox County election night cyberattack was smokescreen for another attack," *USA Today Network*, https://www.knoxnews.com/story/news/local/2018/05/17/knox-county-election-cyberattack-smokescreen-another-attack/620921002 (March 28, 2019).

13. "Theresa May Accuses Vladimir Putin of Election Meddling," *BBC News*, Novem-

ber 14, 2017, https://www.bbc.com/news/uk-politics-41973043 (accessed March 27, 2019).

14. Stepan Krevchenko, "Putin Promises 'Decisive' Protection for Ethnic Russians Abroad," *Bloomberg*, October 31, 2018, https://www.bloomberg.com/news/articles/2018-10-31/putin-promises-decisive-protection-for-ethnic-russians-abroad (accessed March 27, 2019).

15. "Russia's Design in the Black Sea: Extending the Buffer Zone," *Center for Strategic and International Studies*, June 28, 2017, https://www.csis.org/analysis/russias-design-black-sea-extending-buffer-zone (accessed March 27, 2019).

16. John L. Mearsheimer, "Why the Ukraine Crisis Is the West's Fault: The Liberal Delusions That Provoked Putin," *Foreign Affairs*, September/October 2014, https://www.foreignaffairs.com/articles/russia-fsu/2014-08-18/why-ukraine-crisis-west-s-fault (accessed March 27, 2019).

17. Anton Troianovski, "Russian Keeps Getting Hit with Sanctions. Do They Make Sense?" *Washington Post*, August 22, 2018, https://www.washingtonpost.com/world/europe/russia-keeps-getting-hit-with-sanctions-do-they-make-a-difference/2018/08/21/f466db1c-a3ec-11e8-ad6f-080770dcddc2_story.html?utm_term=.a7a9dd912e10 (accessed March 27, 2019).

18. David Fromkin, "The Great Game in Asia," *Foreign Affairs*, Spring 1980, https://www.foreignaffairs.com/articles/south-asia/1980-03-01/great-game-asia (accessed March 27, 2019).

19. "Moldova Concerned Over Russian Troop Movements in Breakaway Region," *Radio Free Europe*, June 18, 2018, https://www.rferl.org/a/moldova-russian-troops-movement-transdniester-breakaway-region/29294739.html (accessed March 27, 2019).

20. "Investigation Uncovers Second Russian Montenegro Coup Suspect," *Balkan Insight*, November 22, 2018, http://www.balkaninsight.com/en/article/media-investigation-identifies-montenegro-coup-suspect-11-22-2018 (accessed March 27, 2019).

21. Patrick Tucker, "Vladimir Putin's Busy, Bloody, and Expensive 2019," *Defense One*, December 28, 2018, https://www.defenseone.com/politics/2018/12/vladimir-putins-busy-bloody-and-expensive-2019/153832/ (accessed March 27, 2019).

22. "Report to Congress on U.S. Sanctions on Russia," *USNI News*, January 14, 2019, https://news.usni.org/2019/01/14/report-congress-u-s-sanctions-russia (accessed March 27, 2019); Jordain Carney, "Senate Rejects Effort to Block Trump on Russia Sanctions," *The Hill*, January 16, 2019, https://thehill.com/homenews/senate/425636-senate-rejects-effort-to-block-trump-on-russia-sanctions (accessed March 27, 2019).

23. Kim Zetter. *Countdown to Zero Day: Stuxnet and the Launch of the World's First Digital Weapon*. The Crown Publishing Group, 2015.

24. David E. Sanger, "Obama Order Sped Up Wave of Cyberattacks against Iran," *New York Times*, June 1, 2012, https://www.nytimes.com/2012/06/01/world/middleeast/obama-ordered-wave-of-cyberattacks-against-iran.html (accessed March 27, 2019).

25. David Strain, "Hacking, Cybersecurity, and Asymmetric Threat," *ITPro Today*, January 14, 2017, https://www.itprotoday.com/strategy/hacking-cyber-security-and-asymmetric-threat (accessed March 27, 2019).

26. Joshua Davis, "Hackers Take Down the Most Wired Country in Europe," *Wired*, August 21, 2007, https://www.wired.com/2007/08/ff-estonia/ (accessed March 27, 2019).

27. Joshua Davis, "Hackers Take Down the Most Wired Country in Europe," *Wired*, August 21, 2007, https://www.wired.com/2007/08/ff-estonia/ (accessed March 27, 2019).

28. Scott Stewart, "Hacking: Another Weapon in the Asymmetrical Arsenal," *RealClear Defense*, January 25, 2018, https://www.realcleardefense.com/articles/2018/01/25/hacking_another_weapon_in_the_asymmetrical_arsenal_112959.html (accessed March 27, 2019).

29. "Army says no to more tanks, but Congress insists," *Fox News*, April 28, 2013, https://www.foxnews.com/politics/army-says-no-to-more-tanks-but-congress-insists (accessed March 28, 2019).

30. Dustin Volz and Jim Finkle, "U.S. Indicts Iranians for Hacking Dozens of Banks, New York Dam," *Reuters*, March 24, 2016, https://www.reuters.com/article/us-usa-iran-cyber/u-s-indicts-iranians-for-hacking-dozens-of-banks-new-york-dam-idUSKCN0WQ1JF (accessed March 27, 2019).

31. Deb Reichmann, "U.S. on Lookout for Possible Iran Cyberattacks after Reinstatement of Sanctions," *Insurance Journal*, August 13, 2018, https://www.insurancejournal.com/news/national/2018/08/13/497767.htm (accessed March 27, 2019).

32. Interview with Douglas Lute, November 28, 2018.

CHAPTER 2

1. "Bush or Kerry: The Electoral College Map," *Los Angeles Times*, May 3, 2004, https://www.latimes.com/la-polldatapage-htmlstory.html (March 27, 2019).

2. "Help America Vote Act," *U.S. Election Assistance Commission*, https://www.eac.gov/about/help-america-vote-act/ (accessed March 27, 2019).

3. Brooks Jackson, "Punchcard Ballots Notorious for Inaccuracies," *CNN*, November 15, 2000, http://www.cnn.com/2000/ALLPOLITICS/stories/11/15/jackson.punchcards/ (accessed March 27, 2019).

4. "State Poll Opening and Closing Times (2018)," *Ballotpedia*, https://ballotpedia.org/State_Poll_Opening_and_Closing_Times_(2018) (accessed March 27, 2019).

5. https://www.pewtrusts.org/en/research-and-analysis/blogs/stateline/2016/03/02/aging-voting-machines-cost-local-state-governments

6. "2004 Battleground States," *RealClearPolitics*, https://www.realclearpolitics.com/bush_vs_kerry_sbys.html#mi.

7. Michael Powell and Peter Slevin, "Several Factors Contributed to 'Lost' Votes in Ohio," The Washington Post, December 15, 2004, https://www.washingtonpost.com/archive/politics/2004/12/15/several-factors-con tributed-to-lost-voters-in-ohio/73aefa72-c8e5-4657-9e85-5ec8b2451202/?utm_term=.485dc7ba0a03 (accessed March 27, 2019).

8. Norman Robbins representing The Greater Cleveland Voter Coalition, "Testimony for House Administration Committee," March 21, 2005, https://moritzlaw.osu.edu/electionlaw/docs/050321-robbins.pdf.

CHAPTER 3

1. "Editing 2004 United States Election Voting Controversies," *Web Archive*, https://web.archive.org/web/20080607140724/http://boxer.senate.gov/news/record.cfm?id=230450 (accessed March 27, 2019).

2. Don Gonyea and Jacki Lyden, "100,000 Turn Out for Obama Rally in St. Louis," *NPR*, October 18, 2008, https://www.npr.org/templates/story/story.php?storyId=9586 7827 (accessed March 27, 2019).

3. "Voter 101: What You Need to Know on Election Day," *ABC News*, November 3, 2008, https://abcnews.go.com/GMA/Politics/story?id=6166407 (accessed March 27, 2019).

4. "Jon Carson," *White House of President Barack Obama*, https://obamawhitehouse.ar chives.gov/blog/author/jon-carson (accessed March 27, 2019).

5. Better Philadelphia Elections Coalition, "Keystone Votes Coalition Report Indicates Thousands of Lost or Late Voter Registration Records in 2016," https://seventy.org/ uploads/files/471453302371191889-voter-registration-report-bpec-press-release-6-5-17 -final.pdf.

6. Ibid.

7. "Obama Looking to Turn Indiana Blue," *CNN*, September 16, 2008, http://www .cnn.com/2008/POLITICS/09/16/indiana.battleground/index.html (accessed March 27, 2019).

8. "John McCain: Top 10 Unfortunate Political One-Liners," *Time*, http://content .time.com/time/specials/packages/article/0,28804,1859513_1859526_1859517,00.html (accessed March 27, 2019).

CHAPTER 4

1. Jill Dougherty, "U.S. calls purported sex tape 'doctored' and 'smear campaign,'" CNN, September 24, 2009, http://www.cnn.com/2009/US/09/24/russia.us.sextape/ (March 27, 2019).

2. Matthew Cole and Brian Ross, "U.S. Protests Russian 'Sex Tape' Used to Smear American Diplomat," ABC News, September 23, 2009, https://abcnews.go.com/Blotter/ russian-sex-tape-smear-american-diplomat/story?id=8651311 (March 27, 2019).

3. James Gordon Meek, "The Other Sochi Threat Russian Spies, Mobsters Hacking Your Smartphones," *ABC News*, February 5, 2014, https://abcnews.go.com/Blotter/sochi -threat-russian-spies-mobsters-hacking-smartphones/story?id=22361222 (accessed March 28, 2019). "Smartphones and NSA Spying," *USA Today*, June 20, 2014, http://www.usato day.com/story/tech/columnist/komando/2014/06/20/smartphones-nsa-spying/10548601 (accessed March 28, 2019).

CHAPTER 5

1. Office of the Director of National Intelligence, "Background to 'Assessing Russian Activities and Intentions in Recent US Elections': The Analytic Process and Cyber Incident Attribution," January 6, 2017, https://www.dni.gov/files/documents/ICA_2017_01.pdf.

2. Adam Entous and Greg Miller, "U.S. Intercepts Capture Senior Russian Officials Celebrating Trump Win," *Washington Post*, January 5, 2017, https://www.washingtonpost .com/world/national-security/us-intercepts-capture-senior-russian-officials-celebrating

-trump-win/2017/01/05/d7099406-d355-11e6-9cb0-54ab630851e8_story.html?utm_term=.e7963892e9d1 (accessed March 28, 2019).

3. Olivia Solon and Sabrina Siddiqui, "Russia-backed Facebook posts 'reached 126m Americans' during US election," *The Guardian*, October 30, 2017, https://www.theguardian.com/technology/2017/oct/30/facebook-russia-fake-accounts-126-million (March 27, 2019).

4. Meghan Keneally, "How you can check if you saw Russian ads on Facebook during the 2016 election," ABC News, December 27, 2017, https://abcnews.go.com/Technology/check-russian-ads-facebook-2016-election/story?id=52010115 (accessed March 27, 2019).

5. The entire report is available at "The Intelligence Community Report on Russian Activities in the 2016 Election," *Washington Post*, http://apps.washingtonpost.com/g/page/politics/the-intelligence-community-report-on-russian-activities-in-the-2016-election/2153/ (accessed March 28, 2019).

6. Ryan D. Enos and Anthony Fowler, "Aggregate Effects of Large-Scale Campaigns on Voter Turnout," *Political Science Research and Methods*, Vol. 6, Issue 4, October 2018.

7. Edward-Isaac Dovere, "How Clinton lost Michigan—and blew the election," Politico, December 14, 2016, https://www.politico.com/story/2016/12/michigan-hillary-clinton-trump-232547 (accessed March 27, 2019).

8. Matthew Cole, Richard Esposito, Sam Biddle, and Ryan Grim, "Top-Secret NSA Report Details Russian Hacking Effort Days Before 2016 Election," *The Intercept*, June 5, 2017.

9. Office of the Director of National Intelligence, "Background to 'Assessing Russian Activities and Intentions in Recent US Elections': The Analytic Process and Cyber Incident Attribution," January 6, 2017, https://www.dni.gov/files/documents/ICA_2017_01.pdf.

10. John Solomon and Alison Spann, "FBI Uncovered Russian Bribery Plot before Obama Administration Approved Controversial Nuclear Deal with Moscow," *Hill*, October 17, 2017, https://thehill.com/policy/national-security/355749-fbi-uncovered-russian-bribery-plot-before-obama-administration (accessed March 28, 2019).

11. Eleanor Watson, "Homeland Security Official 'Suspects' Russia Targeted All 50 States in 2016," *CBS News*, July 11, 2018, https://www.cbsnews.com/news/homeland-security-official-suspects-russia-targeted-all-50-states-in-2016/ (accessed March 28, 2019).

12. Watson, "Homeland Security Official 'Suspects' Russia Targeted All 50 States in 2016," https://www.cbsnews.com/news/homeland-security-official-suspects-russia-targeted-all-50-states-in-2016/ (accessed March 28, 2019).

13. "2018 Year in Review," Elections Infrastructure ISAC—Center for Internet Security, https://www.cisecurity.org/wp-content/uploads/2019/02/EI-ISAC-2018-YIR.pdf.

14. Jeremy Snow, "Einstein's little bro: Used by most states, Albert guards against malware," Fed Scoop, April 1, 2016, https://www.fedscoop.com/ms-isac-einstein-inspired-program-almost-used-by-every-state/ (accessed March 28, 2019).

15. "2018 Year in Review," Elections Infrastructure ISAC—Center for Internet Security, https://www.cisecurity.org/wp-content/uploads/2019/02/EI-ISAC-2018-YIR.pdf.

16. Christopher Bing, "More U.S. States Deploy Technology to Track Election Hacking Attempts," *Reuters*, August 16, 2018, https://www.reuters.com/article/us-usa-election-cy

ber/more-u-s-states-deploy-technology-to-track-election-hacking-attempts-idUS KBN1L11VD (accessed March 28, 2019).

17. "A Handbook for Elections Infrastructure Security," Center for Internet Security, February 2018, https://www.cisecurity.org/wp-content/uploads/2018/02/CIS-Elections-eBook-15-Feb.pdf.

18. Sean Greene, "Statewide Voter Registration Systems," U.S. Election Assistance Commission, August 31, 2017, https://www.eac.gov/statewide-voter-registration-systems/ (accessed March 27, 2019).

19. "Download the First 5 CIS Controls Guide (Version 6.1)," *Center for Internet Security*, https://learn.cisecurity.org/first-five-controls-download.

20. "Election 2016 Recount: Where Five States Stand," *USA Today*, December 6, 2016, https://www.usatoday.com/story/news/nation-now/2016/12/06/election-2016-recount-where-5-states-stand/95052042/ (accessed March 28, 2019).

21. Secretary of State's Office, "Secretary Wyman: Charges of 'Rigged' U.S. Elections Irresponsible," *Washington Secretary of State Blog*, October 18, 2016, https://blogs.sos.wa.gov/fromourcorner/index.php/2016/10/secretary-wyman-charges-of-rigged-us-elections-irresponsible/ (accessed March 28, 2019).

22. "Read the Mueller Report: Searchable Document and Index," *The New York Times*, April 18, 2019, https://www.nytimes.com/interactive/2019/04/18/us/politics/mueller-report-document.html (accessed April 30, 2019).

23. "Zero Days (2016)," *IMDb*, https://www.imdb.com/title/tt5446858/.

24. "Read the Mueller Report: Searchable Document and Index," *The New York Times*, April 18, 2019, https://www.nytimes.com/interactive/2019/04/18/us/politics/mueller-report-document.html (accessed April 30, 2019).

25. Catherine Zhu, "New defense bill bans the U.S. government from using Huawei and ZTE tech, " *TechCrunch*, August 13, 2018, https://techcrunch.com/2018/08/13/new-defense-bill-bans-the-u-s-g overnment-fro m-using-huawei-and-zte-tech; accessed March 28, 2019.

26. "Foreign Economic Espionage in Cyber Space," *National Counterintelligence and Security Center*, July 24, 2018, https://www.dni.gov/files/NCSC/documents/news/20180724-economic-espionage-pub.pdf (accessed March 28, 2019).

27. Rachel Weiner, "Trial Exposes Connections between Cybercriminals and Russian Government," *Washington Post*, May 21, 2018, https://www.washingtonpost.com/local/public-safety/trial-exposes-connections-between-cybercriminals-and-russian-government/2018/05/21/b252268c-584c-11e8-858f-12becb4d6067_story.html?utm_term=.13b71170fb39 (accessed March 28, 2019).

28. "Read the Mueller Report: Searchable Document and Index," *The New York Times*, April 18, 2019, https://www.nytimes.com/interactive/2019/04/18/us/politics/mueller-report-document.html (accessed April 30, 2019).

29. Andy Greenberg, "How an Entire Nation Became Russia's Test Lab for Cyberwar," *Wired*, June 6, 2017, https://www.wired.com/story/russian-hackers-attack-ukraine/ (accessed March 27, 2019).

30. A few examples: Weiner, "Trial Exposes Connections between Cybercriminals and Russian Government," https://www.washingtonpost.com/local/public-safety/trial-expo

ses-connections-between-cybercriminals-and-russian-government/2018/05/21/b252268c-584c-11e8-858f-12becb4d6067_story.html?utm_term=.13b71170fb39.

31. New Jersey and Pennsylvania: J. B. Wogan, "Votes Miscounted? Your State May Not Be Able to Find Out," *Governing.com*, December 2, 2016, http://www.governing.com/topics/politics/gov-states-vote-election-audits-recounts.html (accessed March 28, 2019). Georgia: Kristina Torres, "An Election Primer on Georgia's Election System and Ballot Security," *Atlanta Journal-Constitution*, September 9, 2016, https://www.ajc.com/news/state—regional-govt—politics/election-primer-georgia-voting-system-and-ballot-security/yedbpzowTMxdeBOwjHlkZP/ (accessed March 28, 2019).

32. Sam Levine, "Hackers Tried to Breach a Tennessee County Server on Election Night: Report," *Huffington Post*, May 11, 2018, https://www.huffingtonpost.com/entry/knox-county-election-cyberattack_us_5af5ca21e4b032b10bfa56ee (accessed March 28, 2019).

33. Andy Greenberg, "How an Entire Nation Became Russia's Test Lab for Cyberwar," *Wired*, June 20, 2017, https://www.wired.com/story/russian-hackers-attack-ukraine/ (accessed March 28, 2019).

34. According to former Ukrainian presidential candidate Viktor Yushchenko, the goal is to "destabilize the situation in Ukraine, to make its government look incompetent and vulnerable" (Greenberg, "How an Entire Nation Became Russia's Test Lab for Cyberwar," https://www.wired.com/story/russian-hackers-attack-ukraine/).

CHAPTER 6

1. "Frequently Asked Questions about DEF CON," *DEF CON*, https://www.defcon.org/html/links/dc-faq/dc-faq.html (accessed March 29, 2019).

2. "DEF CON Events Las Vegas 2019," *DEF CON*, https://aivillage.org/events/vegas2018 (accessed March 29, 2019);"DEF CON Biohacking Village," *DEF CON*, https://www.defconbiohackingvillage.org/ (accessed March 29, 2019).

3. "Frequently Asked Questions about DEF CON," https://www.defcon.org/html/links/dc-faq/dc-faq.html.

4. Although the movie script does not explicitly mention DEF CON, ("Sneakers," *Daily Script*, http://www.dailyscript.com/scripts/Sneakers.pdf [accessed March 29, 2019]), a man with with the exact same job as Redford's character spoke at DEF CON soon after the movie came out: "DEF CON 19: Steal Everything, Kill Everyone, Cause Total Financial Ruin!" *YouTube*, https://www.youtube.com/watch?v=JsVtHqICeKE (accessed March 29, 2019).

5. JPat Brown, "FBI Files on DEF CON Show 'Spot the Fed' Contest a Sore Spot for Feds," *Muckrock*, May 6, 2015, https://www.muckrock.com/news/archives/2015/may/06/def-cons-spot-fed-contest-sore-spot-feds/ (accessed March 29, 2019).

6. "Getting to Know Former CIA and FBI Director William Webster," *Capitol File*, https://capitolfile-magazine.com/getting-to-know-former-cia-and-fbi-director-william-webster (accessed March 29, 2019).

7. Adrianne Jeffries, "This Guy Hunted Wi-Fi Hackers Using a Giant Backpack Made Out of Radios," *Outline*, July 31, 2017, https://theoutline.com/post/2017/this-guy

-hunted-wi-fi-hackers-using-a-giant-backpack-made-out-of-radios?zd = 1&zi = vmxfid33 (accessed March 29, 2019).

8. "Frequently asked questions about DEF CON," *DEF CON*, https://www.defcon.org/html/links/dc-faq/dc-faq.html.

9. DEF CON, now one of the largest hacking conferences in the world, began in 1993, just two years after CERN made the World Wide Web available to the public and the same year they announced that access would be free, two actions that were key in opening the internet for public use ("Frequently Asked Questions about DEF CON," https://www.defcon.org/html/links/dc-about.html; Martin Bryant, "20 Years Ago Today, the World Wide Web Opened to the Public," *NextWeb*, August 6, 2011, https://thenextweb.com/insider/2011/08/06/20-years-ago-today-the-world-wide-web-opened-to-the-public/ [accessed March 29, 2019]).

10. "Eunet," *Revolvy*, https://www.revolvy.com/page/EUnet (accessed March 29, 2019).

11. "ARCHIVED: What was BITNET, and what happened to it?," Indiana University, January 18, 2018, https://kb.iu.edu/d/aaso

12. "McDonnell Douglas Corporation," *Encyclopedia Britannica*, https://www.britannica.com/topic/McDonnell-Douglas-Corporation (accessed March 29, 2019).

13. Joel M. Snyder, "Technological Reflections: The Absorption of Data Communications in the Soviet Union," *Opus*, http://www.opus1.com/www/jms/diss.html (accessed March 29, 2019).

14. "Digital Equipment Corporation," *Encyclopedia Britannica*, https://www.britannica.com/topic/Digital-Equipment-Corporati on (accessed March 29, 2019).

15. "Digital Equipment Corporation," https://www.britannica.com/topic/Digital-Equipment-Corporation.

16. "Hacker Ethics," *Chaos Computer Club*, https://www.ccc.de/en/hackerethics (accessed March 29, 2019).

17. Jed Harris, "Nabbed on the Data Highway," *New York Times*, November 26, 1989, http://movies2.nytimes.com/books/99/01/03/specials/stoll-egg.html (accessed March 29, 2019).

18. "BITNET," *Encyclopedia Britannica*, https://www.brittannica.com/technology/BITNET (accessed March 29, 2019); "What Was BITNET, and What Happened to It?" *Indiana University*, January 18, 2018, https://kb.iu.edu/d/aaso (accessed March 29, 2019).

19. "The Origins of Tymnet," *Cap Lore*, http://cap-lore.com/Tymnet/ETH.html (accessed March 29, 2019).

20. "What Is an Internet Service Provider (ISP)?" *Datapath*, https://datapath.io/resources/blog/what-is-an-internet-service-provider/ (accessed March 29, 2019); "Tier 1 Internet Service Provider (Tier 1 ISP)," *Techopedia*, https://www.techopedia.com/definition/23819/tier-1-internet-service-provider-tier-1-isp (accessed March 29, 2019).

21. "About IETF," *IETF.org*, https://www.ietf.org/about/ (accessed March 29, 2019).

22. "An Act of Courage on the Soviet Internet," Slate, August 19, 2016, https://slate.com/technology/2016/08/the-1991-soviet-internet-helped-stop-a-coup-and-spread-a-message-of-freedom.html (accessed March 29, 2019).

23. Soldatov and Borogan, "An Act of Courage on the Soviet Internet," https://slate

.com/technology/2016/08/the-1991-soviet-internet-helped-stop-a-coup-and-spread-a-message-of-freedom.html.

24. The full quote is: "Since this crisis seems to be fading away, I want to thank everyone! See you again, in the next world crisis!" ("Report USSR Gorbachev," *Ibiblio*, https://www.ibiblio.org/pub/academic/communications/logs/report-ussr-gorbatchev [accessed March 29, 2019]).

25. "CIA Told to Focus on Economic Spying," *United Press International*, July 23, 1995, https://www.upi.com/Archives/1995/07/23/Report-CIA-told-to-focus-on-economic-spying/3937806472000/ [accessed March 29, 2019]; David E. Sanger and Tim Weiner, "Emerging Role for the CIA: Economic Spy," *New York Times*, October 15, 1995, https://www.nytimes.com/1995/10/15/world/emerging-role-for-the-cia-economic-spy.html (accessed March 29, 2019).

CHAPTER 7

1. Amy Chozick, "When Clinton Joined Obama Administration, Friction Was Over Staff, Not Email," *The New York Times*, March 5, 2015, https://www.nytimes.com/2015/03/06/us/politics/when-hillary-clinton-joined-obama-administration-friction-was-over-staff-not-email.html (accessed March 27, 2019).

2. Former President Barack Obama and Former Gov. Mitt Romney Debate Transcript, Commission on Presidential Debates, October 22, 2012, https://www.debates.org/voter-education/debate-transcripts/october-22-2012-the-third-obama-romney-presidential-debate/.

3. Jeremy Herb, "Mook Suggests Russians Leaked DNC E-mails to Help Trump," *Politico*, July 24, 2016, https://www.politico.com/story/2016/07/robby-mook-russians-emails-trump-226084 (accessed March 29, 2019).

4. "The Campaign as a Start-Up: A Conversation with Hillary Clinton's Digital Strategy Team," *YouTube*, June 15, 2016, https://www.youtube.com/watch?v=7uJyaL8ikuA (accessed March 29, 2019).

5. "Donald Trump to Russia: Hack and Publish Hillary Clinton's 'Missing' E-Mails," *Guardian*, July 27, 2016, https://www.theguardian.com/us-news/2016/jul/27/donald-trump-russia-hillary-clinton-emails-dnc-hack (accessed March 29, 2019).

6. Andy Greenberg, "How an Entire Nation Became Russia's Test Lab for Cyberwar," *Wired*, June 20, 2017, https://www.wired.com/story/russian-hackers-attack-ukraine/ (accessed March 29, 2019).

7. "Read the Mueller Report," *The New York Times*, April 18, 2019, https://www.nytimes.com/interactive/2019/04/18/us/politics/mueller-report-document.html (accessed April 30, 2019).

8. Illinois shut down their voter database based on fear it had been hacked in July (Pam Fessler, "Hacking an Election: Why It's Not as Far-Fetched as You Might Think," *NPR*, August 1, 2016, https://www.npr.org/2016/08/01/488264073/hacking-an-election-why-its-not-as-far-fetched-as-you-might-think [accessed March 29, 2019]).

9. "Provisional Ballots: November 2, 2004," *Ohio Secretary of State*, https://www.sos

.state.oh.us/elections/election-results-and-data/2004-elections-results/provisional-ballots-november-2-2004/#gref (accessed March 29, 2019).

10. Joe Uchill, "Investigation Shows DHS Did Not Hack Georgia Computers," *Hill*, June 27, 2017, https://thehill.com/policy/cybersecurity/339734-investigation-shows-dhs-did-not-hack-georgia-state-computers (accessed March 29, 2019); Eric Geller, "Elections Security: Federal Help or Power Grab?" *Politico*, August 28, 2016, https://www.politico.com/story/2016/08/election-cyber-security-georgia-227475 (accessed March 29, 2019).

11. Mark Landler and Helene Cooper, "After a Bitter Campaign, Forging an Alliance," *New York Times*, March 18, 2010, https://www.nytimes.com/2010/03/19/us/politics/19policy.html?hp&mtrref=swampland.time.com&gwh=3BF4B6B7B33DA996FF8B90FDE152A0E6&gwt=pay (accessed March 29, 2019).

CHAPTER 8

1. Collin Brennan, "Election 2016 Recount: Where Five States Stand," *USA Today*, December 6, 2016, https://www.usatoday.com/story/news/nation-now/2016/12/06/election-2016-recount-where-5-states-stand/95052042/ (accessed March 30, 2019).

2. Chris Good, "How hackable are American voting machines? It depends who you ask," ABC News, October 15, 2018, https://abcnews.go.com/Politics/hackable-american-voting-machines-depends/story?id=58511054 (accessed March 30, 2019).

3. "NASS Statement on DEFCON Voting Machine Hacking Events," *National Association of Secretaries of State*, August 9, 2018, https://www.nass.org/node/1511 (accessed March 30, 2019).

4. "Testing and Certification Manual, Version 2.0," *Election Assistance Commission*, May 31, 2015, https://www.eac.gov/assets/1/28/Cert.Manual.4.1.15.FINAL.pdf (accessed March 30, 2019).

5. See note 4.

6. "NASS Statement on DEFCON Voting Machine Hacking Events," National Association of Secretaries of State, August 9, 2018, https://www.nass.org/node/1511 (accessed March 29, 2019).

7. "DEFCON 11 Speakers," https://www.defcon.org/html/defcon-11/defcon-11-speakers.html (accessed March 30, 2019).

8. Eric Schmitt, David E. Sanger, and Maggie Haberman, "In Push for 2020 Election Security, Top Official Was Warned: Don't Tell Trump," *The New York Times*, April 24, 2019, https://www.nytimes.com/2019/04/24/us/politics/russia-2020-election-trump.html (accessed April 30, 2019).

9. DEFCON Conference, "DEF CON 25 Voting Village—Jake Braun—Securing the Election Office A Local Response," Youtube, October 18, 2017, https://www.youtube.com/watch?v=v5UcuY33Hic&list=PL9fPq3eQfaaCmT5TovZejSlrBIfaf_epF&index=1

CHAPTER 9

1. Jordan Robertson and Michael Riley, "The Big Hack: How China Used a Tiny Chip to Infiltrate U.S. Companies," Bloomberg, October 4, 2018, https://www.bloomberg

.com/news/features/2018-10-04/the-big-hack-how-china-used-a-tiny-chip-to-infiltrate-america-s-top-companies (accessed March 29, 2019).

2. Jasmin Boyce, "Bloomberg stands by China microchip article as denials and skepticism mount," NBC News, October 23, 2018, https://www.nbcnews.com/tech/tech-news/bloomberg-stands-microchip-stories-denials-skepticism-mount-n923466 (accessed March 30, 2019).

3. "Foreign Economic Espionage in Cyberspace," *National Counterintelligence and Security Center,* July 24, 2018, https://www.dni.gov/files/NCSC/documents/news/20180724-economic-espionage-pub.pdf (accessed March 30, 2019).

4. "About NASS," National Association of Secretaries of State, https://www.nass.org/about-nass (accessed March 30, 2019).

5. "NASS Resolution on Principles for Federal Assistance Funding of Elections," National Association of Secretaries of State, Adopted February 4, 2019, https://www.nass.org/node/1557.

6. "Corporate Affiliates," National Association of Secretaries of State, https://www.nass.org/membership/corporate-affiliates (accessed March 30, 2019).

7. "Polling Place Equipment, November 2018," *Verified Voting*, https://www.verifiedvoting.org/verifier/ (accessed March 30, 2019).

8. Andy Greenberg, "How an Entire Nation Became Russia's Test Lab for Cyberwar," *Wired*, June 20, 2017, https://www.wired.com/story/russian-hackers-attack-ukraine/ (accessed March 30, 2019).

9. Greenberg, "How an Entire Nation Became Russia's Test Lab for Cyberwar," https://www.wired.com/story/russian-hackers-attack-ukraine/.

10. Adam Thorp, "Illinois Election Officials: 'Very Likely' State Was Target of Russian Hackers," *Chicago Sun-Times*, July 13, 2018, https://chicago.suntimes.com/news/illinois-election-officials-very-likely-state-was-target-of-russian-hackers/ (accessed March 30, 2019).

11. Maya T. Prabhu, "Georgia Moves to Secure Federal Grant to Improve Election System," *Atlanta Journal-Constitution*, July 10, 2018, https://www.ajc.com/news/state—regional-govt—politics/georgia-moves-secure-federal-grant-improve-election-system/BwL4GG6DycCMDXeLKdbDEK/ (accessed March 30, 2019).

12. Glenn Chapman, "Hackers School Next Generation at DEF CON Kids," *Phys.org*, June 26, 2011, https://phys.org/news/2011–06-hackers-school-defcon-kids.html (accessed March 30, 2019).

13. Laura Hautala, "Don't Baby These Kid Hackers," *CNET*, August 7, 2016, https://www.cnet.com/news/dont-baby-these-kid-hackers-defcon/ (accessed March 30, 2019).

14. J.M. Porup, "What is sql injection? How SQLi attacks work and how to prevent them," CSO Online, October 2, 2018, https://www.csoonline.com/article/3257429/what-is-sql-injection-how-sqli-attacks-work-and-how-to-prevent-them.html (accessed March 30, 2019).

CHAPTER 10

1. Fred Miller, "Political Naturalism," Stanford Encyclopedia of Philosophy, 2017, https://plato.stanford.edu/entries/aristotle-politics/supplement3.html.

2. "Ian Bassin," *Huffington Post*, https://www.huffingtonpost.com/author/xian-bassinx (accessed April 1, 2019).

3. Seth Rosenblatt, "NSA leaders to hackers: Cybersecurity's a team sport," The Parallax, August 24, 2018, https://the-parallax.com/2018/08/24/nsa-defcon-cybersecurity-team-sport/ (accessed March 30, 2019

4. U.S. Senate, "Russian Targeting of Election Infrastructure During the 2016 Election: Summary of Initial Findings and Recommendations," May 8, 2018, https://www.intelligence.senate.gov/sites/default/files/publications/RussRptInstlmt1.pdf.

5. @mattblaze, "I don't know how to reliably secure a complex internet facing service against a state adversary. No one I know does, either. The only people I'd trust to try understand this," Twitter, August 22, 2018, https://twitter.com/mattblaze/status/1032469778412253189.

6. @mattblaze, "Note: if I comment that it's extremely unlikely that your state or local government election website can withstand attack from a foreign intelligence agency, I'm not casting aspersions on your competence. I'm stating a harsh reality that you really need to think about." Twitter, August 22, 2018,https://twitter.com/mattblaze/status/1032469056010563586.

7. "Read the Mueller Report: Searchable Document and Index," *The New York Times*, April 18, 2019, (accessed April 30, 2019

8. See note 7.

9. DEFCON Conference, "DEF CON 26 VOTING VILLAGE—Neal Kelley and Panel—State, Local Perspectives on Election Security," Youtube, November 2, 2018, https://www.youtube.com/watch?v=hyTPB7OsYu8.

10. See note 9.

11. Robert McMillan and Dustin Volz, "Tensions Flare as Hackers Root Out Flaws in Voting Machines," The Wall Street Journal, August 12, 2018, https://www.wsj.com/articles/tensions-flare-as-hackers-root-out-flaws-in-voting-machines-1534078801 (accessed Apeil 30, 2019).

12. @RGB_Lights, "Ignorance of insecurity does not get you security. We need to examine voting machines, SCADA systems, IOT and other important items in our lives. The investigation of these devices by the hacker community is a service, not a threat." *Twitter*, August 28, 2018, https://twitter.com/RGB_Lights/status/1034416417708363776.

13. @EFF, "Responsible election officials and technologists agree, you can't minitage risks you don't know about. Once again @defcon #votingvillage attendees shed needed light on the steps needed to secure the electoral process." Twitter, August 12, 2018, https://twitter.com/EFF/status/1028818942964330497.

14. See note 12.

15. Comments here: https://twitter.com/essvote/status/1033096189149896706 (accessed April 1, 2019).

16. Some responses, including those from Senator Warner and a spokesperson of Senator Harris, are as follows: Derek Hawkins, "The Cybersecurity 202: Lawmakers Discuss Voting Machine Maker's Claim That Spies Benefit from Election Hacking Demos," *Washington Post*, August 28, 2018, https://www.washingtonpost.com/news/powerpost/paloma/the-cybersecurity-202/2018/08/28/the-cybersecurity-202-lawmakers-dismiss-voting-ma

chine-maker-s-claim-that-spies-benefit-from-election-hacking-demos/5b8430ee1b326b3f31919dcb/?utm_term=.41d0aaf836d7 (accessed April 1, 2019).

17. Tim Starks, "What's Next for Postponed Secure Elections Act," *Politico*, August 23, 2018, https://www.politico.com/newsletters/morning-cybersecurity/2018/08/23/whats-next-for-postponed-secure-elections-act-325469 (accessed April 1, 2019).

18. Derek B. Johnson, "Senators Duel Over Audit Requirements in Election Security Bill," *FCW*, August 21, 2018, https://fcw.com/articles/2018/08/21/election-paper-ballots-bill.aspx (accessed April 1, 2019).

19. "About Congresswoman Jackie Speier," *Congresswoman Jackie Speier*, https://speier.house.gov/about (accessed April 1, 2019).

20. The National Academies of Sciences, Engineering, and Medicine, *Securing the Vote: Protecting American Democracy* (Consensus Study Report), 2018, http://sites.nationalacademies.org/pga/stl/voting/index.htm.

EPILOGUE

1. "Technical Standards for Digital Identification Systems," *World Bank Group*, http://id4d.worldbank.org/about-us; http://documents.worldbank.org/curated/en/707151536126464867/pdf/129743-WP-PUBLIC-ID4D-Catalog-of-Technical-Standards.pdf (accessed April 1, 2019); "ID4D: Identification for Development," *World Bank Group*, https://id4d.worldbank.org/sites/id4d.worldbank.org/files/ID4D%20English%20Flyer_web%2012052018.pdf (accessed April 1, 2019).

2. Kamakshi Ayyar, "The World's Largest Biometric Identification System Survived a Supreme Court Challenge in India," *Time*, September 26, 2018, http://time.com/5388257/india-aadhaar-biometric-identification/ (accessed April 1, 2019)

3. "What Is Open Source?" *Opensource.com*, https://opensource.com/resources/what-open-source (accessed April 1, 2019); "Open Source Definition," *Opensource.org*, https://opensource.org/osd (accessed April 1, 2019).

4. "What Is Open Source?" https://opensource.com/resources/what-open-source.

5. Hilton Collins, "Is Open-Source Software More Secure Than Proprietary Products?" *Government Technology*, July 30, 2009, http://www.govtech.com/security/Is-Open-Source-Software-More-Secure.html (accessed April 1, 2019).

6. "World Bank: What We Do," *World Bank*, http://www.worldbank.org/en/about/what-we-do (accessed April 1, 2019).

7. Stephen Mayhew, "History of Biometrics," *BiometricUpdate.com*, https://www.biometricupdate.com/201802/history-of-biometrics-2 (accessed April 1, 2019).

8. "Mandatory National IDs and Biometric Databases," Electronic Frontier Foundation, https://www.eff.org/issues/national-ids

9. "IDEMIA Delivers the First Smart National Identity Card to Citizens of Nepal," *IDEMIA*, December 12, 2018, https://www.idemia.com/press-release/idemia-delivers-first-smart-national-identity-card-citizens-nepal-2018-12-12 (accessed April 1, 2019); Ananda Gautam, "First Digital ID Card Issued in Panchthar," *Kathmandu Post*, November 19, 2018, http://kathmandupost.ekantipur.com/news/2018-11-19/centenarian-bhandari-acquires-first-digital-national-id-20181119201216.html (accessed April 1, 2019).

Index

About the Author

Jake Braun is executive director of the University of Chicago Harris Cyber Policy Initiative, where he works at the center of politics, technology, and national security to advance the field of cyber policy. He is also cofounder of Cambridge Global Advisors, a national security consulting firm. He has more than 15 years of national security and cybersecurity expertise. Previously, he served in the Obama administration as White House liaison to the U.S. Department of Homeland Security. He is cofounder of the only public third-party inspection of voting equipment in the world, the DEF CON Voting Machine Hacking Village. He has appeared extensively on TV, radio, print, and online media, including CNN, NBC, CBS radio, NPR, CSPAN, *USA Today*, *WIRED* magazine, *Wall Street Journal*, the HBO documentary "Democracy Hacked," and many others. He is co-author of two award-winning reports on election security ("DEF CON 25 Voting Machine Hacking Village: Report on Cyber Vulnerabilities in U.S. Election Equipment, Databases, and Infrastucture" (2017) and DEF CON 26 Voting Machine Hacking Village: Repart on Cyber Vulnerabilities in U.S. Election Equipment, Databases, and Infrastructure" (2018). He teaches at the University of Chicago Harris School of Public Policy and resides in Chicago with his wife, Jena, and two children, Alex and Cordelia, and their dog Athena.